# A Synergistic Framework for Hardware IP Privacy and Integrity Protection

Meng Li • David Z. Pan

# A Synergistic Framework for Hardware IP Privacy and Integrity Protection

Meng Li
Department of Electrical
and Computer Engineering
The University of Texas at Austin
Austin, TX, USA

David Z. Pan
Department of Electrical
and Computer Engineering
The University of Texas at Austin
Austin, TX, USA

ISBN 978-3-030-41249-4 ISBN 978-3-030-41247-0 (eBook)
https://doi.org/10.1007/978-3-030-41247-0



This Springer imprint is published by the registered company Springer Nature Switzerland AG.
The registered company address is: Gewerbestrasse 11, 6330 Cham, Switzerland

# Preface

As the technology node scales down to 45 nm and beyond, the significant increase in design complexity and cost propels the globalization of the $400-billion semiconductor industry. However, such globalization comes at a cost. Although it has helped to reduce the overall cost by the worldwide distribution of integrated circuit (IC) design, fabrication, and deployment, it also introduces ever-increasing intellectual property (IP) privacy and integrity infringement. Recently, primary violations, including stealth hardware Trojan, unauthorized reverse engineering, and malicious fault attacks, have been reported by leading semiconductor companies and resulted in billions of dollars loss annually.

While hardware IP protection strategies are highly demanded, the researches were just initiated lately and still remain preliminary. Firstly, the lack of the mathematical abstractions for these IP violations makes it difficult to formally evaluate and guarantee the effectiveness of the protections. Secondly, the poor scalability and cost-effectiveness of the state-of-the-art protection strategies make them impractical for real-world applications. Moreover, the absence of a holistic IP protection further diminishes the chance to address these highly correlated IP violations which exploit physical clues throughout the whole IC design flow.

To protect hardware IP privacy and integrity, the book proposes a synergistic framework with a focus on security-aware design, optimization, and evaluation. The proposed framework consists of five interacting components that directly target at the primary IP violations. First, to prevent the insertion of the hardware Trojan, a split manufacturing strategy is proposed that achieves formal security guarantee while minimizing the introduced overhead. Then, to hinder reverse engineering, a fast security evaluation algorithm and a provably secure IC camouflaging strategy are proposed. Meanwhile, to impede the fault attacks, a new security primitive, named as public physical unclonable function (PPUF), is designed as an alternative to the existing cryptographic modules. A novel cross-level fault attack evaluation procedure also is proposed to help designers identify security-critical components to protect general purpose processors and compare different security enhancement strategies against the fault attack. All the five algorithms are developed based on rigorous mathematical modeling for primary IP violations and focus on different stages of IC design, which can be combined synergistically to provide a formal security guarantee.

We are particularly grateful to Dr. Meng Li's PhD dissertation committee members, as the major material of this book is based on his dissertation. In particular, we want to thank Prof. Yier Jin for his detailed technical suggestions and guidance. It was a great pleasure to work with him and his group on the exciting area and projects on hardware security. We also want to thank Prof. Mohit Tiwari for his great course on the security issues in hardware/software interface and his insightful suggestions on this research projects. We would also like to thank Prof. Nan Sun and Prof. Nur A. Touba for the helpful discussions and their comments on this dissertation.

We are also grateful to Dr. Rob Aitken (Arm), Dr. Vikas Chandra (Facebook), Dr. Ben Gu (Cadence), Dr. Ru Huang (Peking University), Dr. Liangzhen Lai (Facebook), Dr. Sharad Mehrotra (Cadence), Dr. Jin Miao (Cadence), Dr. Runsheng Wang (Peking University), Dr. Ye Wang (Cadence), Dr. Haoxing Ren (Nvidia), Dr. Naveen Suda (Arm), and Dr. Albert Zeng (Cadence), for their valuable help, suggestions, and discussions on early draft of this book.

We would like to express our gratitude to the colleagues and alumni of the UTDA group at the University of Texas who gave us detailed expert feedback (e.g., Mohamed Baker Alawieh, Shounak Dhar, Wuxi Li, Derong Liu, Yibo Lin, Che-Lun Hsu, Jiaojiao Ou, Biying Xu, Xiaoqing Xu, Wei Ye, Zheng Zhao, Jingyi Zhou, Keren Zhu). Only through those inspiring discussions and productive collaborations that this book could be developed and polished.

We also thank the EDAA and the Springer Press publication team for their help and support in the development of this text. Last but not least, we would like to thank our families for their encouragement and support, as they endured the time demands that writing a book has imposed on us.

Austin, TX, USA Meng Li
Austin, TX, USA David Z. Pan
August 2018

# Contents

**1 Introduction** .......... 1
1.1 Hardware IP Privacy and Integrity Challenges .......... 1
1.2 Overview of This Book .......... 5
References .......... 5

**2 Practical Split Manufacturing Optimization** .......... 9
2.1 Introduction .......... 9
2.2 Preliminary .......... 10
2.2.1 Attack Model of Untrusted Foundries .......... 11
2.2.2 Motivating Example .......... 11
2.2.3 State-of-the-Art Split Manufacturing Flow .......... 12
2.3 Split Manufacturing Security Analysis .......... 13
2.4 *k*-Security Realization .......... 18
2.5 Practical Framework for Trojan Prevention .......... 21
2.5.1 MILP-Based FEOL Generation .......... 21
2.5.2 Lagrangian Relaxation Algorithm .......... 25
2.5.3 *k*-Secure Layout Refinement .......... 29
2.6 Experimental Results .......... 31
2.6.1 Experimental Setup .......... 31
2.6.2 FEOL Generation Strategy Comparison .......... 31
2.6.3 Physical Synthesis Comparison .......... 34
2.6.4 Physical Proximity Examination .......... 35
2.6.5 Relation Between Overhead and Framework Parameters .......... 36
2.7 Summary .......... 37
References .......... 37

**3 IC Camouflaging Optimization and Evaluation** .......... 39
3.1 Introduction .......... 39
3.2 "Arms Race" Evolution .......... 40
3.3 Provably Secure IC Camouflaging .......... 43
3.3.1 Preliminary: Active Learning .......... 44
3.3.2 IC Camouflaging Security Analysis .......... 45

3.3.3 Novel Camouflaging Cell Design ... 48
3.3.4 AND-Tree Camouflaging Strategy ... 52
3.3.5 Provably Secure IC Camouflaging ... 58
3.3.6 Experimental Results ... 68
3.3.7 Summary ... 74
3.4 De-camouflaging Timing-Based Logic Obfuscation ... 75
3.4.1 Preliminary: Timing-Based Camouflaging ... 76
3.4.2 A Motivating Example ... 78
3.4.3 TimingSAT Framework ... 79
3.4.4 Experimental Results ... 87
3.4.5 Summary ... 94
References ... 94

**4 Fault Attack Protection and Evaluation** ... 97
4.1 Introduction ... 97
4.2 Practical PPUF Design ... 97
4.2.1 Preliminaries ... 99
4.2.2 PPUF Topology and ESG Analysis ... 103
4.2.3 PPUF Physical Realization ... 110
4.2.4 Experimental Results ... 112
4.2.5 Summary ... 115
4.3 Cross-Level Monte Carlo Evaluation Framework ... 116
4.3.1 Motivation ... 117
4.3.2 Problem Formulation ... 118
4.3.3 Importance Sampling via System Pre-characterization ... 122
4.3.4 Cross-Level Fault Propagation Simulation ... 126
4.3.5 Experimental Results ... 128
4.3.6 Summary ... 131
References ... 132

**5 Conclusion and Future Work** ... 135

**Index** ... 137

# Acronyms

| | |
|---|---|
| BEOL | Back end of line |
| CRP | Challenge–reponse pair |
| ESG | Evaluation simulation gap |
| FEOL | Front end of line |
| FSM | Finite state machine |
| HD | Hamming distance |
| IC | Integrated circuit |
| IP | Intellectual property |
| KL | Kullback–Leibler |
| KNN | K-nearest neighbor |
| LDD | Lightly doped drain |
| LLN | Law of large number |
| LR | Lagrangian relaxation |
| MILP | Mixed-integer linear programming |
| MPU | Memory protection unit |
| PAC | Probably approximately correct |
| PPUF | Public physical unclonable function |
| PUF | Physical unclonable function |
| RBF | Radial basis function |
| RTL | Register-transfer level |
| SAT | Satisfiability |
| SD | Source degradation |
| SoC | System on chip |
| SPS | Signal probability skew |
| SSF | System security factor |
| SVM | Support vector machine |
| TU | Transformation unit |

# Chapter 1
# Introduction

## 1.1 Hardware IP Privacy and Integrity Challenges

Recent decades have witnessed the significant advances of the whole semiconductor industry. Following Moore's law, the minimum feature size of the semiconductor devices has been shrinking beyond 10 nm. While it enables ICs to achieve higher frequency, better computation capability, and lower energy consumption, it also introduces ever-increasing design and fabrication challenges [2, 5, 9, 33, 33], including a more complicated design flow, an explosion of design corners, emerging design and fabrication constraints, and so on. All the challenges eventually result in a much higher cost for each semiconductor company and propel the whole industry to become more and more globalized. As shown in Fig. 1.1, in the modern supply chain, IC design, fabrication, testing, assembly, and deployment are distributed worldwide, with different companies from a variety of countries involved and undertaking different tasks.

The globalized supply chain significantly reduces the cost for the semiconductor companies and shortens the time to market of major products. However, it also introduces more and more severe hardware IP violations [4, 10, 23, 29, 31, 36, 38]. As shown in Fig. 1.2, in the globalized supply chain, the design houses take full charge of the hardware design from the register-transfer level (RTL), gate level, to physical level. Then, the IC designs are shipped across the world for fabrication, packaging, and development. Due to the lack of control over the supply chain after the shipment of the designs, violations of hardware IP privacy and integrity emerge.

Primary IP violations can be categorized into three classes: hardware Trojan [1, 15, 28, 34], reverse engineering [14, 26, 27, 29], and fault attack [3, 11, 31, 38], as shown in Fig. 1.2. Hardware Trojans are malicious modifications made by untrusted foundries. For example in Fig. 1.3 (c), gate 6 is inserted maliciously at the output of gate 2. The detection of such malicious gate can be challenging for uninformed designers if the trigger signal $t$ is deliberately selected. To prevent Trojan insertion, split manufacturing is proposed to hide the critical signals, including the trigger



M. Li, D. Z. Pan, *A Synergistic Framework for Hardware IP Privacy and Integrity Protection*, https://doi.org/10.1007/978-3-030-41247-0_1

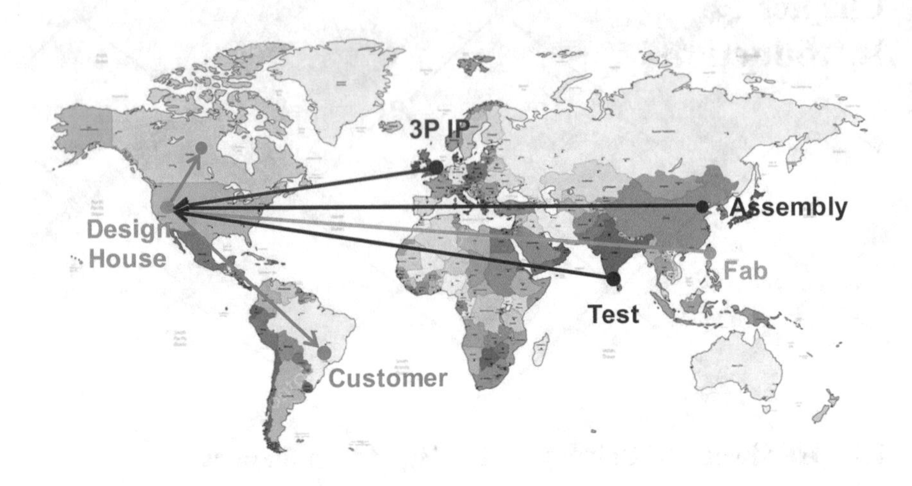


Fig. 1.1 Globalized IC supply chain

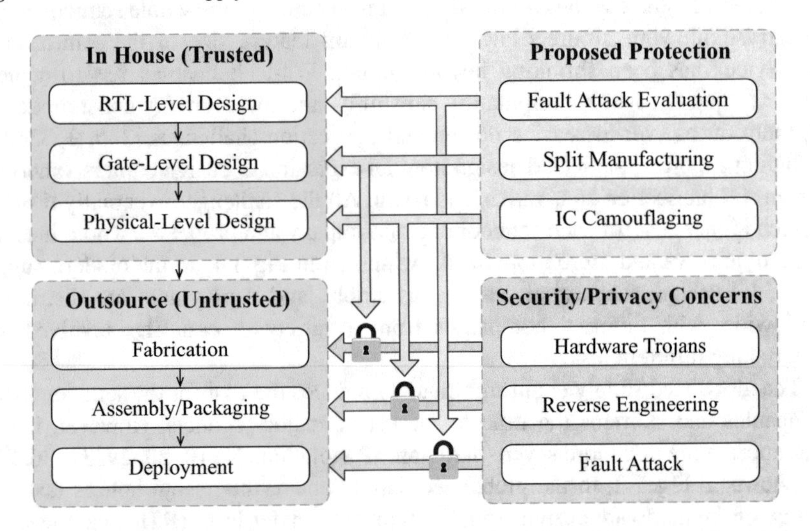


Fig. 1.2 Security concerns of the modern globalized supply chain and the proposed protections

signals, e.g., signal $t$, and the target signals, e.g., node 2 [7, 13, 22, 32]. The key idea is to split a circuit layout into the front-end-of-line (FEOL) layers and back-end-of-line (BEOL) layers as shown in Fig. 1.3 (a). BEOL layers only consist of the wires in higher metal layers and are fabricated in-house, through which critical signals can be protected. For example, consider the FEOL and BEOL layers shown in Fig. 1.3 (d) and (e). By hiding the wires in the BEOL layers, the attackers cannot distinguish gate $B$ and $D$, and thus, cannot determine which gate in the FEOL layers implements node 2.

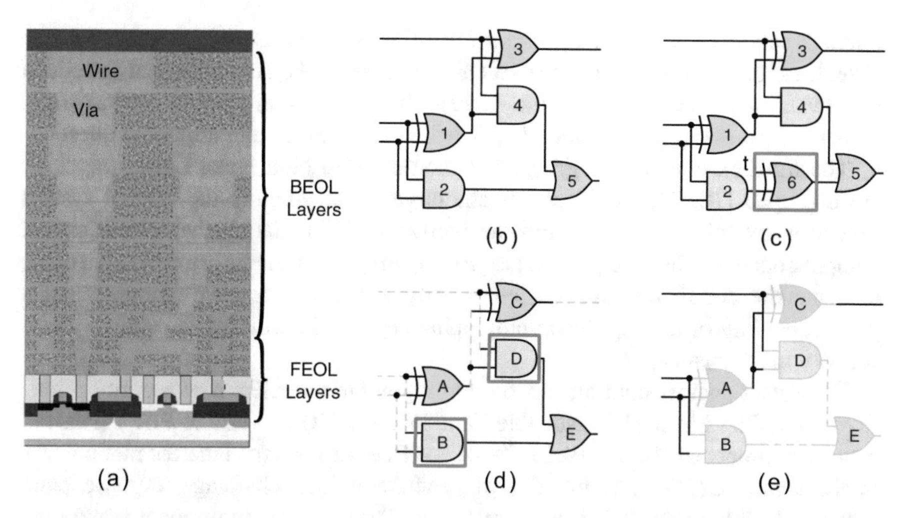


**Fig. 1.3** Example of split manufacturing: (**a**) definition of the FEOL and BEOL layers [8], (**b**) the original netlist, (**c**) Trojan inserted netlist, (**d**) the FEOL layers of the design, and (**e**) the BEOL of the design

Although split manufacturing demonstrates promising protection against Trojan insertion, a critical problem is that the security cannot be guaranteed for all the signals simply by hiding the BEOL layers. For example in Fig. 1.3, regardless of the wires hidden in the BEOL layers, node 5 can always be identified as it has a different gate type compared with the other nodes. Besides the lack of guarantee on the security, the state-of-the-art algorithms also suffer from poor scalability. This mainly originates from the repetitive security evaluation, which requires computation-intensive operations on large graphs [7].

The second class of IP violations comes from reverse engineering. After the fabrication, the chips are sent out for assembly, packaging and eventually go to the open market. The attackers can now strip the chip layer by layer to reconstruct the gate-level netlist and re-distribute it without the authorization from the IP vendors. To prevent reverse engineering, IC camouflaging emerges as a remedy [6, 20, 35, 37, 40]. By leveraging fabrication-level techniques, camouflaging cells are first designed to implement different functionalities, even though their layouts look the same to the attackers. They are then inserted into the circuit for obfuscating the circuit functionality.

Extensive researches have been conducted to insert the camouflaging cells so that the security can be maximized under the overhead constraints [12, 35, 37]. Meanwhile, different attack strategies are also proposed to resolve the correct circuit functionality [25]. Although the "arms race" between the attack and protection inspires better defense against reverse engineering, the key shortcoming is that due to the lack of a formal security metric, the security of all the camouflaging strategies is only evaluated with the existing attack methods. The empirical nature of such evaluation leads to an overestimation of the security level. Meanwhile, an efficient security evaluation algorithm for different camouflaging strategies is also absent, especially considering the recent advances in the strategies that introduce unconventional structures into the netlist [40]. Therefore, how to evaluate and enhance the security of the camouflaged netlist becomes another important question.

The third class of IP integrity issues comes from fault attacks after the chips are deployed [18, 30, 39]. Fault attacks target at obstructing the normal system execution by injecting errors into the hardware. By radiating the critical circuit components with high energy particle strikes, voltage transients are created in order to make the circuit malfunction temporarily. Fault attacks have demonstrated a great capability of leaking the cryptographic keys, and nullifying the entire system security mechanisms [18, 30, 39].

To protect against fault attacks on cryptographic modules, new security primitives, including Physical Unclonable Functions (PUFs) and Public PUFs (PPUFs) have been proposed [16, 17, 19, 21, 24, 41]. A PUF exploits the inherent randomness in the scaled CMOS technologies to generate unique challenge-response pairs (CRPs). A PPUF is a PUF that is created so that its simulation model is publicly available but large discrepancies exist between the execution delay and simulation time [19, 21]. While a PPUF relies on the time gap between execution and simulation (ESG) to derive this security, the ESG of all the existing PPUFs remains unjustified in terms of theoretical soundness and physical practicality. Theoretical soundness requires the ESG to be bounded rigorously, especially considering the advanced parallel and approximate computing schemes. Physical practicality further requires that the ESG can be realized effectively considering the existing fabrication techniques and the generated output must be measurable. The unjustified ESG significantly limits the application of the PPUFs in modern designs.

To hinder fault attacks on general purpose processors, an accurate evaluation of the system vulnerability is highly demanded to identify critical circuit components and guide the design optimization. However, it is challenging to quantitatively evaluate the system vulnerability due to the probabilistic nature of the fault attack process. The randomness of the attack process mainly comes from the uncertainty of fault injection techniques. Our study on commercial processors reveals that failing to account for the attack uncertainty can result in over-pessimistic estimation of the system vulnerability. Nonetheless, it is not easy to capture the attack uncertainty efficiently. Though statistical metrics have been proposed [18, 39], an explicit enumeration of the system state space is usually required for an accurate evaluation, whose complexity increases exponentially with the system size.

Thus far, the IP violations described above are already hard to protect against by themselves. However, they can be combined together to further empower the attackers. One example is that fault attacks can help to trigger the inserted hardware Trojans and the Trojans, in turn, provide fault attacks with a much more fine-grained and accurate attack technique. The serious hardware IP violations call for a synergistic protection.

## 1.2 Overview of This Book

This book proposes a synergistic framework for hardware IP privacy and integrity protection that helps the IP vendors from design, optimization, and evaluation perspectives.

Chapter 2 presents a practical split manufacturing algorithm. For the first time, the insertion of dummy gates and wires is exploited in the split manufacturing process. The existing security definition is extended to accommodate the inserted dummy gates and wires. A sufficient condition is also derived to avoid the computationally intensive security evaluation when generating the FEOL layers, which can be realized with fast relaxation algorithms and achieves 1400× speedup compared with the state-of-the-arts.

Chapter 3 proposes a provably secure IC camouflaging technique against the satisfiability-based (SAT-based) reverse engineering attack. The equivalence between the attack and the Boolean function learning problem is built for the first time, based on which, a quantitative security metric is derived. The key security impacting factors are identified with different techniques proposed to achieve an exponential increase of the security level with a linear increase of overhead. Meanwhile, a novel attack strategy is also proposed for the security evaluation, named as TimingSAT. TimingSAT is able to evaluate all the existing camouflaging techniques and demonstrates a great efficiency.

Chapter 4 proposes a fast yet accurate fault attack evaluation flow and a novel PPUF design to hinder the fault attack on general purpose processors and cryptographic modules. In the evaluation flow, a probabilistic model is developed and a statistical metric is proposed to capture the attack uncertainty. A Monte Carlo strategy is also proposed to evaluate the statistical metric efficiently, which is further empowered with an importance sampling strategy. For the proposed PPUF design, its execution is equivalent to solving the hard-to-parallel and hard-to-approximate max-flow problem in a complete graph on the chip. Thus, the max-flow problem can be used as the simulation model to bound the ESG rigorously. To enable an efficient physical realization, a crossbar structure is developed and the source degeneration technique is exploited to map the graph topology on the chip.

Chapter 5 summarizes this book and discusses the future research directions.

## References

1. Bhasin, S., & Regazzoni, F. (2015). A survey on hardware Trojan detection techniques. In *Proceedings of the IEEE International Symposium on Circuits and Systems* (pp. 2021–2024).
2. Borkar, S. (1999). Design challenges of technology scaling. *IEEE Micro, 19*(4), 23–29.
3. Chen, C.-N., & Yen, S.-M. (2003). Differential fault analysis on AES key schedule and some countermeasures. In *Proceedings of the Australasian Conference on Information Security and Privacy*.
4. Chen, S., Chen, J., Forte, D., Di, J., Tehranipoor, M., & Wang, L. (2015). Chip-level anti-reverse engineering using transformable interconnects. In *Proceedings of the IEEE*

*International Symposium on Defect and Fault Tolerance in VLSI and Nanotechnology Systems* (pp. 109–114).
5. De, V., & Borkar, S. (1999). Technology and design challenges for low power and high performance. In *Proceedings of the International Symposium on Low Power Electronics and Design* (pp. 163–168). New York: ACM.
6. El Massad, M., Garg, S., & Tripunitara, M. V. (2015). Integrated circuit (IC) decamouflaging: Reverse engineering camouflaged ICs within minutes. In *Proceedings of the Network and Distributed System Security Symposium.*
7. Imeson, F., Emtenan, A., Garg, S., & Tripunitara, M. (2013). Securing computer hardware using 3D integrated circuit (IC) technology and split manufacturing for obfuscation. In *USENIX Security Symposium* (pp. 495–510). Berkeley: USENIX.
8. International Technology Roadmap for Semiconductor. (2014). Available: https://www.itrs.net/. Accessed Nov 2014.
9. Jhaveri, T., Rovner, V., Liebmann, L., Pileggi, L., Strojwas, A. J., & Hibbeler, J. D. (2010). Co-optimization of circuits, layout and lithography for predictive technology scaling beyond gratings. *IEEE Transactions on Computer-Aided Design of Integrated Circuits and Systems, 29*(4), 509–527.
10. Krieg, C., Wolf, C., & Jantsch, A. (2016). Malicious LUT: A stealthy FPGA Trojan injected and triggered by the design flow. In *Proceedings of the International Conference on Computer Aided Design.*
11. Lemke-Rust, K., & Paar, C. (2006). An adversarial model for fault analysis against low-cost cryptographic devices. In *Proceedings of the IEEE Workshop Fault Diagnosis and Tolerance in Cryptography.*
12. Li, M., Shamsi, K., Meade, T., Zhao, Z., Yu, B., Jin, Y., et al. (2016). Provably secure camouflaging strategy for IC protection. In *Proceedings of the International Conference On Computer Aided Design* (pp. 28:1–28:8).
13. Li, M., Yu, B., Lin, Y., Xu, X., Li, W., & Pan, D. Z. (2018). A practical split manufacturing framework for Trojan prevention via simultaneous wire lifting and cell insertion. In *Proceedings of the Asia and South Pacific Design Automation Conference* (pp. 265–270).
14. Li, W., Gascon, A., Subramanyan, P., Tan, W. Y., Tiwari, A., Malik, S., et al. (2013). WordRev: Finding word-level structures in a sea of bit-level gates. In *Proceedings of the IEEE International Symposium on Hardware Oriented Security and Trust* (pp. 67–74).
15. Liu, Y., Jin, Y., & Makris, Y. (2013). Hardware Trojans in wireless cryptographic ICs: Silicon demonstration & detection method evaluation. In *2013 IEEE/ACM International Conference on Computer-Aided Design (ICCAD)* (pp. 399–404).
16. Miao, J., Li, M., Roy, S., Ma, Y., & Yu, B. (2017). SD-PUF: Spliced digital physical unclonable function. *IEEE Transactions on Computer-Aided Design of Integrated Circuits and Systems, 37*, 927–940.
17. Miao, J., Li, M., Roy, S., & Yu, B. (2016). LRR-DPUF: Learning resilient and reliable digital physical unclonable function. In *Proceedings of the International Conference on Computer-Aided Design.*
18. Nahiyan, A., Xiao, K., Yang, K., Jin, Y., Forte, D., & Tehranipoor, M. (2016). AVFSM: A framework for identifying and mitigating vulnerabilities in FSMs. In *Proceedings of the IEEE/ACM Design Automation Conference.*
19. Potkonjak, M., & Goudar, V. (2014). Public physical unclonable functions. *Proceedings of the IEEE, 102*(8), 1142–1156.
20. Rajendran, J., Sam, M., Sinanoglu, O., & Karri, R. (2013). Security analysis of integrated circuit camouflaging. In *Proceedings of the 2013 ACM SIGSAC Conference on Computer & Communications Security* (pp. 709–720).
21. Rührmair, U. (2009). SIMPL systems: On a public key variant of physical unclonable functions. IACR Cryptology ePrint Archive, 2009, 255.
22. Shi, Q., Xiao, K., Forte, D., & Tehranipoor, M. M. (2017). Securing split manufactured ICs with wire lifting obfuscated built-in self-authentication. In *Proceedings of the IEEE Great Lakes Symposium on VLSI* (pp. 339–344).

23. Skorobogatov, S., & Woods, C. (2012). Breakthrough silicon scanning discovers backdoor in military chip. In *Proceedings of the International Conference on Cryptographic Hardware and Embedded Systems*.
24. Sreedhar, A., & Kundu, S. (2011). Physically unclonable functions for embedded security based on lithographic variation. In *Proceedings of the Design, Automation and Test in Europe* (pp. 1–6).
25. Subramanyan, P., Ray, S., & Malik, S. (2015). Evaluating the security of logic encryption algorithms. In *Proceedings of the IEEE International Symposium on Hardware Oriented Security and Trust* (pp. 137–143).
26. Subramanyan, P., Tsiskaridze, N., Li, W., Gascon, A., Tan, W. Y., Tiwari, A., et al. (2014). Reverse engineering digital circuits using structural and functional analyses. *IEEE Transactions on Emerging Topics in Computing, 2*(1), 63–80.
27. Sugawara, T., Suzuki, D., Fujii, R., Tawa, S., Hori, R., Shiozaki, M., et al. (2014). Reversing stealthy dopant-level circuits. In *Proceedings of the International Conference on Cryptographic Hardware and Embedded Systems* (pp. 112–126). Berlin: Springer.
28. Tehranipoor, M., & Koushanfar, F. (2010). A survey of hardware Trojan taxonomy and detection. *IEEE Design & Test of Computers, 27*(1), 10–25.
29. Torrance, R., & James, D. (2009). The state-of-the-art in IC reverse engineering. In *Proceedings of the International Conference on Cryptographic Hardware and Embedded Systems* (pp. 363–381). Berlin: Springer.
30. Tunstall, M., Mukhopadhyay, D., & Ali, S. (2011). Differential fault analysis of the advanced encryption standard using a single fault. In *Proceedings of the International Workshop on Information Security Theory and Practices*.
31. Van Woudenberg, J. G., Witteman, M. F., & Menarini, F. (2011). Practical optical fault injection on secure microcontrollers. In *Proceedings of the IEEE Workshop Fault Diagnosis and Tolerance in Cryptography*.
32. Wang, Y., Chen, P., Hu, J., & Rajendran, J. J. (2017). Routing perturbation for enhanced security in split manufacturing. In *Proceedings of the Asia and South Pacific Design Automation Conference* (pp. 605–510).
33. Warnock, J. (2011). Circuit design challenges at the 14nm technology node. In *Proceedings of the IEEE/ACM Design Automation Conference* (pp. 464–467). New York: ACM.
34. Xiao, K., Forte, D., Jin, Y., Karri, R., Bhunia, S., & Tehranipoor, M. M. (2016). Hardware Trojans: Lessons learned after one decade of research. *ACM Transactions on Design Automation of Electronic Systems (TODAES), 22*, 1–23.
35. Xie, Y., & Srivastava, A. (2016). Mitigating SAT attack on logic locking. In *Proceedings of the International Conference on Cryptographic Hardware and Embedded Systems* (pp. 127–146).
36. Yang, K., Hicks, M., Dong, Q., Austin, T., & Sylvester, D. (2016). A2: Analog malicious hardware. In *Proceedings of the IEEE Symposium on Security and Privacy*.
37. Yasin, M., Mazumdar, B., Sinanoglu, O., & Rajendran, J. (2016). CamoPerturb: Secure IC camouflaging for minterm protection. In *Proceedings of the International Conference on Computer Aided Design* (pp. 29:1–29:8).
38. Yuce, B., Ghalaty, N. F., Deshpande, C., Patrick, C., Nazhandali, L., & Schaumont, P. (2016). FAME: Fault-attack aware microprocessor extensions for hardware fault detection and software fault response. In *Proceedings of the International Workshop on Hardware and Architectural Support for Security and Privacy*.
39. Yuce, B., Ghalaty, N. F., & Schaumont, P. (2015). TVVF: Estimating the vulnerability of hardware cryptosystems against timing violation attacks. In *Proceedings of the IEEE International Symposium on Hardware Oriented Security and Trust*.
40. Zhang, L., Li, B., Yu, B., Pan, D. Z., & Schlichtmann, U. (2018). TimingCamouflage: Improving circuit security against counterfeiting by unconventional timing. In *Proceedings of the Design, Automation and Test in Europe*.
41. Zheng, Y., Hashemian, M. S., & Bhunia, S. (2013). RESP: A robust physical unclonable function retrofitted into embedded SRAM array. In *Proceedings of the IEEE/ACM Design Automation Conference* (pp. 60:1–60:9).

# Chapter 2
# Practical Split Manufacturing Optimization

## 2.1 Introduction

Hardware Trojans inserted by untrusted foundries directly threaten the whole system security. While extremely harmful, the detection of such hardware Trojans can be very difficult because they are inserted during fabrication time and activated under a very restricted condition, which allows the Trojans to bypass both the pre- and post-silicon testing phases. Therefore, how to prevent the Trojan insertion by untrusted foundries is becoming a key issue.

To prevent Trojan insertion proactively, split manufacturing is proposed [5, 8, 9, 15, 18, 20–22, 27, 28]. In the split manufacturing process, the circuit layout is split into FEOL layers, which consist of all the cells and interconnections in lower metal layers, and BEOL layers, which consist of all the interconnections in higher metal layers. Because the fabrication of BEOL layers usually requires less advanced technologies, it is affordable to maintain such trusted foundries for the BEOL layer fabrication, by which important circuit information can be hidden to prevent Trojan insertions by untrusted foundries.

In recent years, different split manufacturing frameworks have been proposed. The first formal security criterion for split manufacturing against Trojan insertion, named as $k$-security, is proposed in [9]. A circuit is defined to be $k$-secure if for each cell in the original netlist, there exist $k$ cells in the FEOL layers that can be its actual physical implementation and are indistinguishable to the attackers. The security definition is formalized based on graph isomorphism [3], as will be discussed in Sect. 2.3. To realize $k$-security, a greedy algorithm is also proposed to determine the wires to be lifted from the FEOL layers to the BEOL layers. In [4, 13, 17, 23–25], techniques in physical synthesis stage, including fault-analysis based pin swapping, placement perturbation, and so on, are proposed to prevent the untrusted foundries from reverse engineering the hardware IP. These methods are proposed under another orthogonal attack model, the main target of which is



M. Li, D. Z. Pan, *A Synergistic Framework for Hardware IP Privacy and Integrity Protection*, https://doi.org/10.1007/978-3-030-41247-0_2

to prevent reverse engineering by untrusted foundries. It is currently not clear how these proposed methods can be leveraged for hardware Trojan prevention.

Despite the extensive researches on split manufacturing, existing approaches still suffer from insufficient security guarantee, poor computational efficiency, and large performance overhead as will be detailed in Sect. 2.2. In this chapter, besides wire lifting, the insertion of dummy cells and wires are considered simultaneously to address the security and practicality issues of existing methods. Considering existing security criterion cannot model the situation where FEOL layers contain cells and wires that do not exist in the original netlist, we propose a new criterion that is fully compatible with the insertion of dummy nodes and wires. Our security criterion can also balance the trade-off between security and overhead by allowing the flexibility of protecting any arbitrary subset of circuit nodes. We further derive a sufficient condition for the security criterion to avoid the computationally intensive graph isomorphism checking and enable an efficient security realization. To realize the security criterion while minimizing the introduced overhead, we propose a holistic framework. Our framework consists of a novel mixed-integer linear programming (MILP) formulation for the FEOL layer generation and a Lagrangian relaxation (LR) algorithm [12, 14] to significantly speedup the generation process. A layout refinement technique is also proposed to guarantee security in the physical synthesis stage. We summarize our contributions as follows:

- A new security criterion fully compatible with cell and wire insertion is proposed with its sufficient condition derived to enable an efficient split manufacturing process.
- An MILP-based formulation is proposed to generate the FEOL layers considering dummy cell and wire insertion with wire lifting simultaneously and further accelerated with an LR-based algorithm.
- A layout refinement technique is proposed to guarantee security in the physical synthesis stage.
- The proposed flow is validated by extensive experimental results and demonstrates good efficiency and practicality.

The rest of the chapter is organized as follows. Section 2.2 defines the attack model and describes an example to illustrate the motivation of the chapter and the state-of-the-art split manufacturing flow in detail. Section 2.3 formally formulates the split manufacturing problem and defines our new security criterion. Section 2.4 proposes a sufficient condition to achieve the proposed criterion. Section 2.5 describes our split manufacturing framework. Section 2.6 demonstrates the performance of the framework, followed by conclusion in Sect. 2.7.

## 2.2 Preliminary

In this section, the attack model of untrusted foundries is first reviewed. A motivating example is analyzed to explain the insufficiency when only wire lifting is considered in the split manufacturing flow. We also describe the state-of-the-art FEOL generation flow proposed in [9] in detail.

### 2.2.1 Attack Model of Untrusted Foundries

We consider attackers from untrusted foundries that target at inserting malicious hardware Trojans into the design. We assume the following attack model as described in [9]:

- The attacker has the gate-level netlist of the design.
- The attacker has full knowledge of the FEOL layers, including the cells and wires in lower metal layers as well as their physical information.
- The attacker knows the algorithms of generating the FEOL layers but does not know the specific mapping between the cells in the FEOL layers and the original netlist.

The assumption on the knowledge of the gate-level netlist is pretty strong but indeed possible. The main reason is that the attackers who intend for such Trojan insertion can potentially be resourceful enough to have malicious observers in the design stage [9]. Meanwhile, the profit of a successful Trojan insertion can also be pretty large, especially for military applications [19]. Given the gate-level netlist, the attackers can first determine the target gates in the design for the Trojan insertion. Then, the attackers will try to identify the physical implementation of the target gates based on the information of the FEOL layers and insert the Trojan.

### 2.2.2 Motivating Example

As described in Sect. 2.2.1, given the information on the original circuit netlist and the FEOL layers, the attackers can try to locate the actual implementation for the target gates identified in the original netlist. According to [9], the attack process can be formulated as searching for a bijective mapping of gates in the FEOL layers to the gates in the original netlist. Consider the circuit netlist as shown in Fig. 2.1a and the FEOL layers shown in Fig. 2.1b. There exist 4 distinct bijective mappings between the FEOL layers and the original netlist, i.e. $f_1 : \{1, 2, 3, 4, 5\} \rightarrow \{1', 2', 3', 4', 5'\}$, $f_2 : \{1, 2, 3, 4, 5\} \rightarrow \{1', 3', 2', 4', 5'\}$, $f_3 : \{1, 2, 3, 4, 5\} \rightarrow \{1', 2', 3', 5', 4'\}$, $f_4 : \{1, 2, 3, 4, 5\} \rightarrow \{1', 3', 2', 5', 4'\}$. Following the current mapping relations, both Gates $2'$ and $3'$ in the FEOL layers can be mapped to Gate 2. From the attacker's perspective, both Gates $2'$ and $3'$ can implement Gate 2 in the original netlist. Therefore, if the attacker targets at Gate 2 for the Trojan insertion, his capability to accurately insert the Trojan is significantly weakened.

However, there are at least two problems with the FEOL layers in Fig. 2.1b. On the one hand, for Gate 1, only $1'$ in the FEOL layers shares the same functionality,

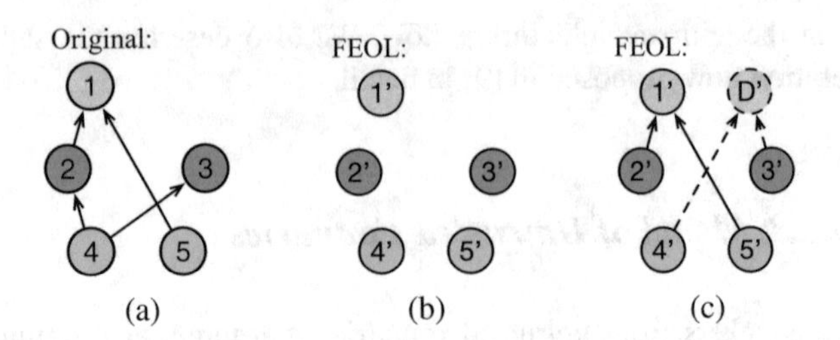


**Fig. 2.1** A motivating example of split manufacturing process and the insufficiency of state-of-the-art framework: (**a**) the original netlist; (**b**) FEOL layers generated by the original flow; (**c**) FEOL layers generated by our new framework (nodes with the same colors have the same functionalities and the dotted lines indicate the inserted dummy edges)

which indicates the attacker can always determine its identity. In fact, because the other gates in the original netlist all have different functionalities compared to Gate 1, simply by lifting wires to the BEOL layers can never help enhance the security of Gate 1. On the other hand, in Fig. 2.1b, all the wires are lifted to the BEOL layers. Because there are usually much fewer routing resources in higher metal layers, design houses are forced to either increase the number of layers fabricated in trusted foundries or reduce the area utilization to mitigate the routing congestion in higher metal layers, both of which increase the overhead of split manufacturing significantly.

In this chapter, we propose a new framework that considers dummy gate and wire insertion simultaneously with the wire lifting. Consider the FEOL layers shown in Fig. 2.1c. A dummy gate $D'$ of the same gate type as 1 is inserted. Two dummy wires $(3', D')$ and $(4', D')$ are inserted to the FEOL layers as well. In this way, for any gate targeted by the attacker in the original circuit, there are two gates that cannot be distinguished in the FEOL layers. Meanwhile, only two wires, i.e. $(4', 2')$ and $(4', 3')$, are lifted to the BEOL layers while the number of wires in the FEOL layers remain the same as that in the original netlist. Thereby, the two drawbacks of the original framework [9] can be well solved.

However, it should be noted that due to the insertion of dummy gates and wires, the bijective mappings between the original netlist and the FEOL layers do not hold anymore, which means the original formalization of the attack process and the definition of security criterion cannot be applied anymore. In Sect. 2.3, we will propose our new formulation for the split manufacturing protection and the Trojan insertion attack.

### 2.2.3 *State-of-the-Art Split Manufacturing Flow*

In Sect. 2.2.2, we use a motivating example to compare the FEOL layers generated by the original flow and our framework. In this section, we will review the process

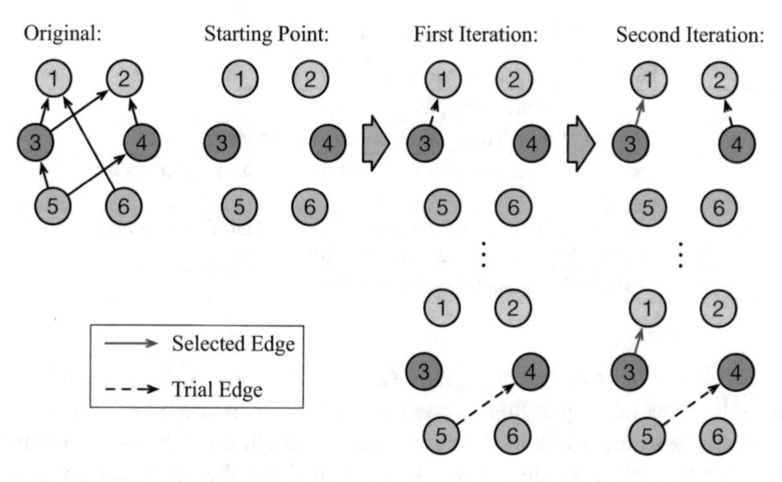


**Fig. 2.2** Traditional split manufacturing flow

of the FEOL layer generation proposed in [9]. Consider the original netlist shown in Fig. 2.2. To determine the wires to be lifted to the BEOL layers, the proposed framework starts by lifting all the wires to the BEOL layers first. Then, it adds the wires back to the FEOL layers iteratively following a greedy selection strategy. In each iteration, it tries to add each wire back to the FEOL layers, and then determine the security level for the current FEOL layers. The wire that provides the best security level will be selected and added back to the FEOL layers. The procedure continues until the security level can no longer be satisfied.

The state-of-the-art split manufacturing flow suffers from scalability issue. As described above, in each iteration, to determine the security level when a wire is added back, repetitive checking is carried out to search for the bijective mappings between the whole circuit and the FEOL layers. Although it can be elegantly formulated as a SAT problem, the computation cost makes the method intractable quickly even for small benchmark circuits. In this chapter, we target at solving all the above-mentioned problems of the existing method to provide better security guarantee, reduce the introduced overhead, and enhance the scalability of the split manufacturing flow.

## 2.3 Split Manufacturing Security Analysis

In this section, we will formulate the split manufacturing problem as a graph problem. To accommodate the insertion of dummy cells and wires, we will formally define the split manufacturing process and the attack process, and propose a new security criterion. For convenience, some notations used in this chapter are summarized in Table 2.1, which will be defined and explained in detail in this section.

**Table 2.1** Notations used for security definition and analysis

| | |
|---|---|
| $\phi(v)$ | Corresponding node that implements $v$ in the FEOL layers |
| $\ell(v)$ | Label of $v$ representing its cell type |
| $\omega(v)$ | Weight of $v$ indicating whether it is selected for protection |
| $\mathscr{C}(v)$ | Set of candidate nodes in the FEOL layers that can implement $v$ from the attacker's point of view |
| $\mathscr{S}_v(v')$ | Set of spanning subgraph isomorphism relations that map $v'$ in FEOL layers to $v$ in the original netlist |
| $\mathscr{P}_v(v')$ | Probability of candidacy for $v' \in \mathscr{C}(v)$ |

A circuit can be regarded as a graph $G = \langle V, E, \ell, \omega \rangle$. $V$ is the set of vertices, with each vertex corresponding to one circuit node. $E$ is the set of directed edges corresponding to the wires in the circuit. Label function $\ell : V \rightarrow [t]$ maps each vertex to a cell type, where $[t] = \{1, \ldots, t\}$ denotes the set of all possible cell types in the circuit. $\omega : V \rightarrow \{0, 1\}$ assigns a binary weight to each vertex with $\omega(v) = 1$ indicating that the vertex $v$ is selected for protection. $\omega$ is defined to make the framework flexible to protect a subset of circuit nodes[1] due to overhead constraints and balance the trade-off between security and the introduced overhead.

The original netlist and the generated FEOL layers can be represented as two graphs. For the graph representation of the original circuit, denoted as $G$, $V_G$, $E_G$, and $\ell_G$ are straightforward to define. $\omega_G$ is determined by the designer considering the circuit functionality, overhead constraints, and so on. To determine these parameters for the graph representation of the FEOL layer, denoted as $H$, we need to consider its generation process. To generate $H$, for each $v \in V_G$, we add $v'$ to $V_H$ such that $\ell_H(v') = \ell_G(v)$ and $\omega_H(v') = \omega_G(v)$. We denote $v' = \phi(v)$ as the corresponding node for $v$, which represents the actual cell in FEOL that implements $v$ in the netlist. Meanwhile, for each $(v, u) \in E_G$, we add $(\phi(v), \phi(u))$ to $E_H$. Then, we consider the three operations for the generation of $H$:

1. **wire lifting**: if $(u', v') \in E_H$ is lifted to BEOL, then, $E_H = E_H \setminus \{(u', v')\}$ with $V_H, \ell_H$, and $\omega_H$ unchanged;
2. **dummy node insertion**: if $u'$ with the cell type $\ell_{u'}$ is inserted, then, $V_H = V_H \cup \{u'\}$ with $\ell_H(u') = \ell_{u'}, \omega_H(u') = 0$ and $E_H$ is unchanged;
3. **dummy wire insertion**: if $(u', v')$ is inserted, then, $E_H = E_H \cup \{(u', v')\}$ with $V_H, \ell_H$, and $\omega_H$ unchanged.

It should be noted that to guarantee the circuit functionality is not changed and to get rid of floating input pins, we only allow inserting wires pointing to the dummy nodes. Based on the description of the allowed operations, $V_H$, $E_H$, $\ell_H$, and $\omega_H$ can be acquired accordingly.

*Example 2.1* Consider an example of $G$ and $H$ in Fig. 2.3. In $G$, we have nodes 1 and 2 with the same cell type, i.e. $\ell_G(1) = \ell_G(2)$. Assume that we select nodes 1

[1] In the chapter, nodes and cells are the same and used interchangeably.

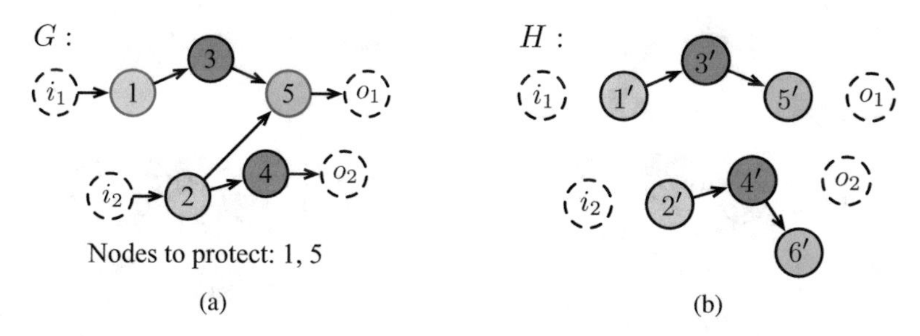


**Fig. 2.3** Example of (**a**) Original graph $G$ (nodes with red stroke have non-zero weights), (**b**) FEOL graph $H$. $i_1$ and $i_2$ are input pins while $o_1$ and $o_2$ are output pins

and 5 for protection, then, $\omega_G(1) = \omega_G(5) = 1$. To generate $H$, we first add the corresponding nodes to $H$ for each node in $G$, i.e. $1', 2', 3', 4', 5'$. Then, we add node $6'$ and wire $(4', 6')$ to $H$ and lift wire $(2', 5')$. Therefore, we have $\omega_H(1') = \omega_G(1) = 1$ and $\omega_H(5') = \omega_G(5) = 1$. For the other nodes in $H$, we have $\omega_H(2') = \omega_H(3') = \omega_H(4') = \omega_H(6') = 0$.

As described in Sect. 2.2, to insert a Trojan, the attacker will first select $v \in V_G$ based on the analysis of the design and then, try to locate its corresponding node $\phi(v)$ in $H$. To formalize the process of locating $\phi(v)$, state-of-the-art method [9] leverages the concept of graph isomorphism.

**Definition 2.1 (Graph Isomorphism)** Two graphs $G_1$ and $G_2$ are isomorphic if there exists a bijective mapping $f : V_{G_1} \rightarrow V_{G_2}$ such that $(u, v) \in E_{G_1}$ if and only if $(f(u), f(v)) \in E_{G_2}$ and $\ell_{G_1}(u) = \ell_{G_2}(f(u)), \ell_{G_1}(v) = \ell_{G_2}(f(v))$.

Because only wire lifting is considered in existing methods, we must have $V_H = V_G$ and $E_H \subseteq E_G$. Therefore, there must be a subgraph of $G$ that is isomorphic to $H$, based on which for each $v \in V_G$, a set of nodes can be identified that may implement $v$ in FEOL. This enables the previous work [9] to formally define the security criterion.

However, *when the insertion of dummy wires and cells are considered, the original isomorphic relation is not satisfied any more*. This is because $H$ contains nodes and edges that do not present in $G$ so that $V_G \neq V_H$ and $E_H \not\subseteq E_G$. To formalize the relation between $G$ and $H$, we first have the following observations on the relations between $H$ and $G$ that always hold

- $\forall v \in V_G, \exists v' \in V_H$ s.t. $v' = \phi(v)$.
- $\forall v', u' \in V_H$, if $\exists u \in V_G$ s.t. $u' = \phi(u)$, then, if $\forall (v', u') \in E_H$, then, there must exist $v \in V_G$ s.t. $v' = \phi(v)$ and $(v, u) \in E_G$.

The first observation indicates that for each circuit node in $G$, there must be one node in $H$ that implements it. The second observation indicates that if $u' \in V_H$ is the corresponding node of $u$ in the netlist, then, for all the edges that points to $u'$, e.g. $(v', u') \in E_H$, there must be $v \in V_G$ with $v'$ as the corresponding node and

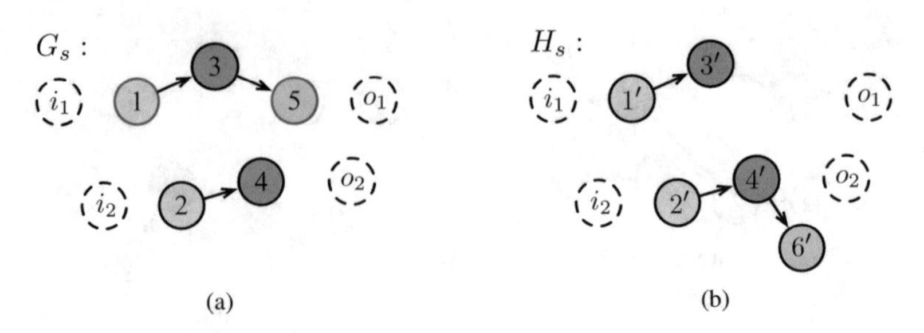


**Fig. 2.4** Example of (**a**) spanning subgraph $G_s$ of $G$ in Fig. 2.3a, and (**b**) induced subgraph $H_s$ of $H$ in Fig. 2.3b

$v$ is connected to $u$ in $G$. This is because we are not allowed to add dummy edges pointing to the corresponding node of $u \in V_G$. For example, in Fig. 2.3, suppose $5' = \phi(5)$, since we are not allowed to add any dummy edges pointing to $5'$, we must be able to find $3 \in V_G$ such that $3' = \phi(3)$ and $(3, 5) \in E_G$. To formalize the relations described above, we leverage the concept of spanning subgraph and induced subgraph [26].

**Definition 2.2 (Spanning Subgraph)** A subgraph $G_s$ of $G$ is referred to as a spanning subgraph if $V_{G_s} = V_G$.

**Definition 2.3 (Induced Subgraph)** A subgraph $G_s$ of $G$ is referred to as an induced subgraph if $\forall(u, v) \in E_G$ with $u, v \in V_G$, $(u, v) \in E_{G_s}$ if and only if $u, v \in V_{G_s}$.

*Example 2.2* Consider an example shown in Fig. 2.4. $G_s$ is a spanning subgraph of $G$ in Fig. 2.3a since $V_{G_s} = V_G$. $H_s$ in Fig. 2.4b is an induced subgraph of $H$ in Fig. 2.3b because for any pair of nodes in $H_s$, if there exists an edge between them in $H$, the edge also exists in $H_s$. For example, nodes $1'$ and $3'$ exist in $H_s$. Because $(1', 3') \in E_H$, for $H_s$ to be an induced subgraph, we must have $(1', 3') \in E_{H_s}$.

Then, considering the spanning subgraph of $G$ and the induced subgraph of $H$, we define the relation of spanning subgraph isomorphism as below.

**Definition 2.4 (Spanning Subgraph Isomorphism)** Given two graphs $G$ and $H$, we say that $G$ is spanning subgraph isomorphic to $H$ if there exists a spanning subgraph of $G$ that is isomorphic to an induced subgraph of $H$.

Spanning subgraph isomorphism defines the criterion for the attackers to identify the corresponding node $\phi(v)$ in FEOL for a target node $v$ in the netlist. For example, in Fig. 2.4, since $G_s$ and $H_s$ are isomorphic, $G$ is spanning subgraph isomorphic to $H$ with $1, 2, 3, 4, 5$ being matched to $2', 1', 4', 3', 6'$, respectively. Therefore, $2'$ is possible to implement node 1 in the final layout from the attacker's point of view. We denote $2'$ as the candidate node for 1.

For the spanning subgraph isomorphism relation, there is one additional constraint to consider. Because inserting dummy wires pointing to the corresponding

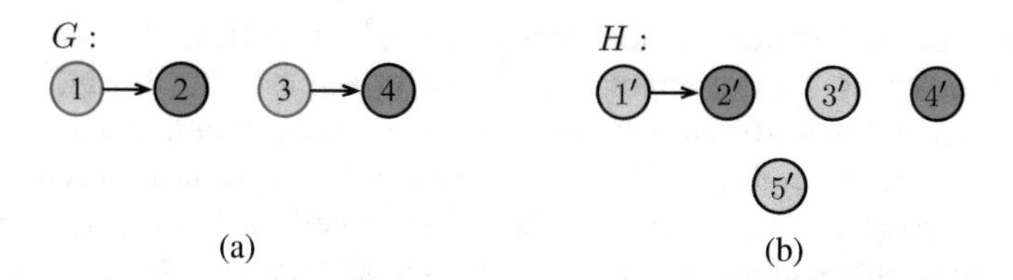


**Fig. 2.5** Example on the weight and probability difference for different candidate nodes: (**a**) the original graph $G$ and (**b**) $H$ is spanning subgraph isomorphic to $G$ with multiple valid isomorphism relations

nodes in the FEOL layers is not allowed, it is possible for some spanning subgraph isomorphism relation to be invalid. For example, consider $G$ and $H$ as shown in Fig. 2.5. There exists a spanning subgraph isomorphism relation that maps $1, 2, 3, 4$ in $G$ to $5', 2', 3', 4'$ in $H$, respectively. Following the current mapping, node $1'$ becomes dummy. However, because $2'$ is the corresponding node of 2 in the current mapping and we are not allowed to insert dummy edges pointing to the corresponding node, $(1', 2')$ must be an edge that exists in the original netlist, which is contradictory to the conjecture that node $1'$ is dummy. We define the spanning subgraph isomorphism relations that satisfy the constraints on wire insertion as valid isomorphism relations. Only the valid isomorphism relations can enhance the security against hardware Trojan insertion.

The proposed spanning subgraph isomorphism relation is more general compared with the graph isomorphism relation. *When only wire lifting is considered, it reduces to the graph isomorphism. It can also capture the situations where $V_G \neq V_H$ and $E_H \nsubseteq E_G$, which enables us to consider cell and wire insertion in the split manufacturing process.*

Because multiple spanning subgraph isomorphism relations may exist between $G$ and $H$, for $v \in V_G$, a set of candidate nodes can be identified, denoted as the candidate set $\mathscr{C}(v)$. For the nodes in the candidate set, the number of spanning subgraph isomorphism relations that can map them to the original node is different. For example, as shown in Fig. 2.5, $1'$, $3'$, and $5'$ are the candidate nodes for 3. For $1'$, there are two different isomorphism relations mapping it to 1, i.e. $f_1 : \{1, 2, 3, 4\} \to \{1', 2', 3', 4'\}$ and $f_2 : \{1, 2, 3, 4\} \to \{1', 2', 5', 4'\}$. For $3'$ and $5'$, there is only one isomorphism relation mapping each of them to 1, i.e. $f_3 : \{1, 2, 3, 4\} \to \{3', 4', 1', 2'\}$ and $f_4 : \{1, 2, 3, 4\} \to \{5', 4', 1', 2'\}$. The nodes with a larger number of spanning subgraph isomorphism relations are more likely to be recognized and selected by the attackers. Therefore, for $v \in V_G$, we define the probability of candidacy for $v' \in \mathscr{C}(v)$ as

$$\mathscr{P}_v(v') = \frac{|\mathscr{S}_v(v')|}{\sum_{u' \in \mathscr{C}(v)} |\mathscr{S}_v(u')|}, \tag{2.1}$$

where $\mathscr{S}_v(v')$ denotes the set of valid spanning subgraph isomorphism relations that maps $v'$ to $v$ and $| \cdot |$ calculates the cardinality of the set.

Besides the difference on the probability of candidacy, the weight of the candidate nodes is also different. As shown in Fig. 2.5, $1'$, $3'$, and $5'$ are the candidate nodes for 3. Because $1'$ and $3'$ are the corresponding nodes of 1 and 3, they have non-zero weights, while for $5'$, the weight is zero since it is dummy.

Now we propose our security criterion for a cell as follows to capture the spanning subgraph isomorphism relation and the observations above.

**Definition 2.5 ($k$-Secure Cell)** Given original graph $G$ and FEOL graph $H$, we say that $v \in V_G$ is $k$-secure with respect to $G$ and $H$ if

$$\sum_{u' \in \mathscr{C}(v)} \mathscr{P}_v(u')\omega_H(u') \leq \frac{1}{k}.$$

Following the definition above, for each $v \in V_G$ with $k$-security, the probability to pick a candidate node with a non-zero weight from $\mathscr{C}(v)$ is limited within $1/k$. In this way, the difference on weight and the probability of candidacy are enforced in the security criterion. Now we define the security criterion for the circuit netlist.

**Definition 2.6 ($k$-Security)** Given $G$ and $H$, we say that $\langle G, H\rangle$ is $k$-secure if $\forall v \in V_G$ with $\omega_G(v) = 1$, $v$ is $k$-secure with respect to $G$ and $H$.

By the above security criterion, we can guarantee that for any node that the attackers may target at, the probability to insert the Trojan into a node with a non-zero weight is always no greater than $1/k$. In this way, by making $k$ large enough, we can guarantee much higher cost and risk for the Trojan insertion.

## 2.4 $k$-Security Realization

To determine the spanning subgraph isomorphism relation, isomorphism checkings between the subgraphs of $G$ and $H$ are usually required, which can be very computation intensive. To avoid direct graph comparison, we adopt recent progress in privacy preserving network publishing [3] to derive a sufficient condition for $k$-security. Our heuristic solution relies on the following concept denoted as $k$-isomorphism [3].

**Definition 2.7 ($k$-Isomorphism)** A graph is $k$-isomorphic if it consists of $k$ disjoint isomorphic subgraphs.

For example, the graph $H$ of FEOL in Fig. 2.3b is 2-isomorphic with $V_{H_{s,0}} = \{1', 3', 5'\}$ and $V_{H_{s,1}} = \{2', 4', 6'\}$. Specifically, we call nodes $1'$ and $2'$ in the same position of $H_{s,0}$ and $H_{s,1}$. For $1'$ and $2'$, if $1' \in \mathscr{C}(1)$, then, $2' \in \mathscr{C}(1)$. Moreover, we must have $\mathscr{P}_1(1') = \mathscr{P}_1(2')$. Assume $1' = \phi(1)$, then, if $\omega(2') = 0$, 1 is 2-secure with respect to $G$ and $H$ in Fig. 2.3. Based on the observation, we have the following lemma for a $k$-isomorphic graph.

**Lemma 2.1** *Given G and $H = \{H_{s,0}, \ldots, H_{s,k-1}\}$, which is k-isomorphic. $\forall v \in V_G$ with $\omega_G(v) = 1$ and $\phi(v) \in V_{H_{s,i}}$, where $i \in \{0, \ldots, k-1\}$, if each $u' \in V_{H_{s,j}} (j \neq i)$, where $u'$ and $\phi(v)$ are in the same position of $H_{s,j}$ and $H_{s,i}$, respectively, satisfies $\omega_H(u') = 0$, then, v is k-secure with respect to G and H.*

***Proof 2.1*** Consider $v \in V_G$ with $\omega_G(v) = 1$ and $H = \{H_{s,0}, \ldots, H_{s,k-1}\}$, which is $k$-isomorphic. Recall $\mathscr{C}(v)$ denotes the candidate set of $v$ and for each $v' \in \mathscr{C}(v)$, the probability of candidacy, i.e. $\mathscr{P}_v(v')$, is defined in Eq. (2.1). For $v' \in \mathscr{C}(v)$, without loss of generality, we assume $v' \in V_{H_{s,0}}$. Then, in $H_{s,1}, \ldots, H_{s,k-1}$, there must be $k-1$ other nodes in the same position as $v'$ that are also in $\mathscr{C}(v)$ and have the same probability of candidacy. Let $\mathscr{L}_i(v)$ be the set of positions of the nodes in $H_{s,i}$ that are in $\mathscr{C}(v)$, for $i \in \{0, \ldots, k-1\}$. Then, we have $\mathscr{L}_0(v) = \ldots = \mathscr{L}_{k-1}(v)$.

Let $V_{H_{s,i}}(j)$ be the node in the $j$th position of $H_{s,i}$, then, from the definition of the probability of candidacy, we have

$$\begin{aligned}
\sum_{v' \in \mathscr{C}(v)} \mathscr{P}_v(v') &= \sum_{i=0}^{k} \sum_{j \in \mathscr{L}_i(v)} \mathscr{P}_v(V_{H_{s,i}}(j)) \\
&= k \sum_{j \in \mathscr{L}_0(v)} \mathscr{P}_v(V_{H_{s,0}}(j)) \\
&= 1.
\end{aligned}$$

Therefore,

$$\sum_{j \in \mathscr{L}_0(v)} \mathscr{P}_v(V_{H_{s,0}}(j)) = \frac{1}{k}.$$

Meanwhile,

$$\begin{aligned}
\sum_{v' \in \mathscr{C}(v)} \mathscr{P}_v(v')\omega_H(v') &= \sum_{i=0}^{k-1} \sum_{j \in \mathscr{L}_i(v)} \mathscr{P}_v(V_{H_{s,i}}(j))\omega_H(V_{H_{s,i}}(j)) \\
&= \sum_{j \in \mathscr{L}_0(v)} \sum_{i=0}^{k-1} \mathscr{P}_v(V_{H_{s,i}}(j))\omega_H(V_{H_{s,i}}(j)) \\
&= \sum_{j \in \mathscr{L}_0(v)} \mathscr{P}_v(V_{H_{s,0}}(j)) \sum_{i=0}^{k-1} \omega_H(V_{H_{s,i}}(j)) \\
&\leq \sum_{j \in \mathscr{L}_0(v)} \mathscr{P}_v(V_{H_{s,0}}(j)) \\
&= \frac{1}{k}.
\end{aligned}$$

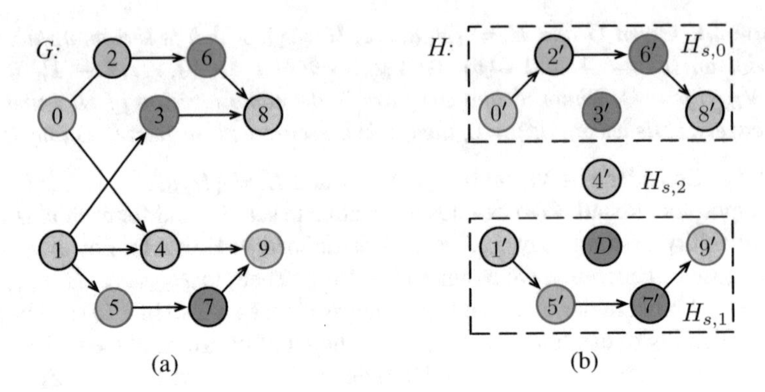


**Fig. 2.6** Example for Theorem 2.1: $G$ is 2-secure with respect to $H$

Note the inequality holds because following Theorem 2.1, for $j$th position in all the $k$ subgraphs, there are at most 1 node with non-zero weight, i.e. $\sum_{i=0}^{k-1} \omega_H(V_{H_{s,i}}(j)) \leq 1$.

Therefore, $v$ is a $k$-secure cell. Because the property holds for all the nodes with non-zero weights, $\langle G, H \rangle$ must be $k$-secure. Hence proved.

Lemma 2.1 formalizes the condition for $v \in V_G$ to be $k$-secure. Because we only require the nodes with non-zero weight to be $k$-secure, we have the following theorem for $k$-security.

**Theorem 2.1** *Given* $G$ *and* $H$, *assume* $H = \{H_{s,0}, \ldots, H_{s,k}\}$, *where* $\{H_{s,0}, \ldots, H_{s,k-1}\}$ *are k-isomorphic. G is k-secure with respect to H if* $\forall v \in V_G$ *with* $\omega_G(v) = 1$, *the following conditions are satisfied:*

1. $\phi(v) \in V_{H_{s,i}}$ *where* $i \in \{0, \ldots, k-1\}$.
2. $\omega_H(u') = 0$, $\forall u' \in V_{H_{s,j}}$ $(j \in \{0, \ldots, k-1\}, j \neq i)$, *where* $u'$ *and* $\phi(v)$ *are in the same position of* $H_{s,j}$ *and* $H_{s,i}$, *respectively.*

*Example 2.3* Consider the example shown in Fig. 2.6. $H$ is composed of 3 subgraphs with $H_{s,0}$ and $H_{s,1}$ being isomorphic to each other. Nodes with non-zero weights like $3', 5', 6', 9'$ are either in $H_{s,0}$ or in $H_{s,1}$, while the weights of the nodes in the same position as them, i.e. $D, 2', 7', 8'$ are zero. Therefore, they are 2-secure with respect to $G$ and $H$ according to Lemma 2.1. Node 4 remains unprotected since its weight is zero. Therefore, $\langle G, H \rangle$ is 2-secure. By introducing weights for each node and $H_{s,k}$, our framework is flexible to protect an arbitrary subset of circuit nodes to balance the trade-off between security and the introduced overhead.

Theorem 2.1 works as a sufficient condition for the proposed security criterion. It is not only fully compatible with the insertion of dummy cells and wires, but also eliminates the requirements and computation overhead of determining the security level through graph isomorphism checkings in the FEOL generation process. The remaining question is how to effectively and efficiently achieve the requirements posed in Theorem 2.1. In the next section, we will describe our split manufacturing flow for the FEOL layer generation.

## 2.5 Practical Framework for Trojan Prevention

In this section, we propose our framework to generate the FEOL and BEOL layers. The inputs to the framework include the original circuit netlist and the selected nodes for protection. An MILP-based formulation, which considers the insertion of dummy wires and gates with wire lifting simultaneously, is first proposed to generate the $k$-secure FEOL layers. We further propose a novel LR-based algorithm and a minimum-cost-flow [6, 10, 12] transformation to enhance the scalability of the framework. In the second step, we propose a layout refinement technique, which enables us to leverage commercial tools for physical synthesis while guarantee the security in the placement stage.

### 2.5.1 *MILP-Based FEOL Generation*

Following the sufficient condition proposed in Theorem 2.1, to achieve $k$-security, we need to generate $H = \{H_{s,0}, \ldots, H_{s,k}\}$ from $G$ so that all the nodes with non-zero weights are added to the first $k$ subgraphs. Because the insertion of dummy wires and nodes is allowed, one trivial solution to generate $H$ is to copy $G$ for $k - 1$ times. *This indicates that k-security can always be achieved when the insertion of dummy cells and wires is considered.* However, such a naive solution usually suffers from large overhead.

To reduce the introduced overhead, in this section, we propose a novel FEOL generation algorithm, whose pseudo code is shown in Algorithm 1. Our algorithm anonymizes all the selected nodes iteratively until all the nodes with non-zero weights are added to $H_{s,0}, \ldots, H_{s,k-1}$. In each iteration, we select $k$ nodes of the

**Algorithm 1** Iterative FEOL generation

1: // $V_r$: the set of nodes that have not been inserted
2: $V_{crit} \leftarrow \{v \in V_G : \omega_G(v) = 1\}$, $V_r \leftarrow V_G$;
3: **while** $V_{crit} \neq \varnothing$ **do**
4: $\quad V_{min} \leftarrow \varnothing$, $c_{min} \leftarrow +\infty$;
5: $\quad$ // $[t]$: the set of cell types
6: $\quad$ **for** $i \in [t]$ **do**
7: $\quad\quad V_i \leftarrow \{v \in V_r : \ell(v) = i\}$;
8: $\quad\quad V_{sel}, c_{sel} \leftarrow$ `NodeSelect`$(k, V_i)$;
9: $\quad\quad$ **if** $c_{min} > c_{sel}$ **then**
10: $\quad\quad\quad V_{min} \leftarrow V_{sel}$, $c_{min} \leftarrow c_{sel}$;
11: $\quad\quad$ **end if**
12: $\quad$ **end for**
13: $\quad$ `InsertToFEOL`$(V_{min}, H_{s,0}, \ldots, H_{s,k-1})$;
14: $\quad V_{crit} \leftarrow V_{crit} \setminus V_{min}$, $V_r \leftarrow V_r \setminus V_{min}$;
15: **end while**
16: $H_{s,k} \leftarrow V_r$;

**Table 2.2** Notations used in the MILP formulation

| | |
|---|---|
| $x_i$ | $x_i = 1$ if the $i$th node is selected |
| $x_{ij}$ | $x_{ij} = 1$ if the $i$th node is inserted to $H_{s,j}$ |
| $\omega_i$ | Weight of the $i$th node |
| $d_j$ | $d_j = 1$ if a dummy node is inserted to $H_{s,j}$ |
| $y_l$ | $y_l = 1$ if an edge can be added from $l$th location to current location in $H_{s,0}, \ldots, H_{s,k-1}$ |
| $y_{lj}$ | $y_{lj} = 1$ if an edge can be added from $l$th location to current location in $H_{s,j}$ |
| $z_l$ | $z_l = 1$ if an edge can be added from current location to $l$th location in $H_{s,0}, \ldots, H_{s,k-1}$ |
| $z_{lj}$ | $z_{lj} = 1$ if an edge can be added from current location to $l$th location in $H_{s,j}$ |
| $IN_{ij}$ | Set of starting locations of edges pointing to current location that can be added if $i$th node is added to $H_{s,j}$ |
| $OUT_{ij}$ | Set of ending locations of edges pointing from current location that can be added if $i$th node is added to $H_{s,j}$ |
| $RES_i$ | Set of edges connected to $i$th node from unadded node |

same label and make sure that exactly one node has a non-zero weight to satisfy Theorem 2.1 (lines 4–10). To select the nodes, we first cluster all the remaining nodes by their labels and then, select $k$ nodes from each cluster with the minimized cost through an MILP-based formulation. The $k$ nodes with the minimized cost among all the clusters are selected and inserted to $H_{s,0}, \ldots, H_{s,k-1}$ (line 11). The iterative algorithm continues until all the nodes with non-zero weights are added to $H_{s,0}, \ldots, H_{s,k-1}$.

The core part of the FEOL generation algorithm is the MILP-based node selection, i.e. `NodeSelect`. Before we introduce our MILP formulation, we list the notations used in the formulation in Table 2.2 and use the following example to illustrate the iterative strategy and the problem that we will solve in each iteration.

*Example 2.4* Consider the original graph $G$ in Fig. 2.6a. Assume nodes $0, 3, 5, 6, 9$ are selected for protection and the required security level is 2. To generate $H$ from $G$, our strategy is to iteratively anonymize the selected nodes with non-zero weights by adding them to $H_{s,0}$ and $H_{s,1}$. As shown in Fig. 2.7a–c, in the first three iterations, nodes $0, 2, 6$ and nodes $1, 5, 7$ are added to $H_{s,0}$ and $H_{s,1}$, respectively. For the nodes in the same position in $H_{s,0}$ and $H_{s,1}$, e.g. 2 and 5 in the first location, only one of them has a non-zero weight, which follows the requirement in Theorem 2.1. In each iteration, to select the nodes to insert into $H_{s,0}$ and $H_{s,1}$, we propose an MILP-based formulation to select a pair of nodes that share the same label and achieve the smallest insertion cost. We use Fig. 2.7c to explain the MILP formulation. Consider node 9 that has a non-zero weight. To anonymize it, we can find nodes 8 and 4 of the same cell type as node 9 and also allow the insertion of dummy nodes $d_0$ and $d_1$. If we add node 4 to $H_{s,0}$, because edge $(0, 4)$ exists in $G$, we have $IN_{40} = \{0\}$, which indicates that there is one edge, i.e. $(0, 4)$, pointing from the 0th location in $H_{s,0}$ to the current location that can be added to $H_{s,0}$ if node 4 is inserted. Similarly, if we add node 9 to $H_{s,1}$, because edge $(7, 9)$ exists in $G$, we have $IN_{91} = \{3\}$. For the

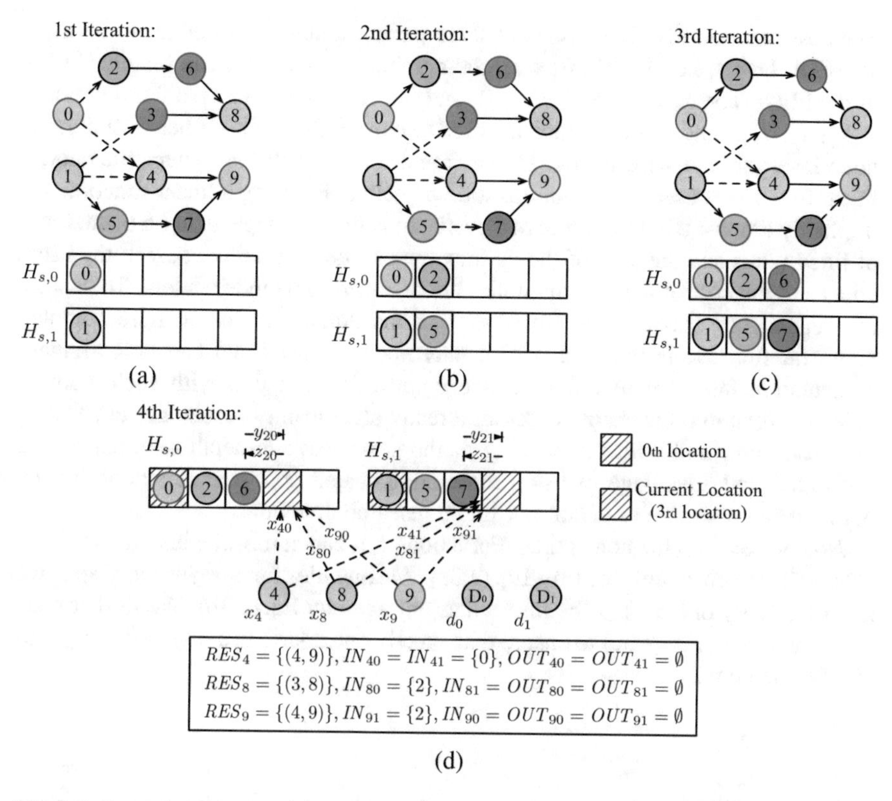


**Fig. 2.7** Example of the iterative strategy and the formulation in each iteration: (**a**)–(**c**) the first three iterations (dotted lines are the wires to be lifted to BEOL layers); (**d**) parameters and formulation for the fourth iteration

dummy nodes $D_0$, we have $IN_{D0} = \{0, 1, 2\}$ and $OUT_{D0} = \emptyset$. This is because to retain the correct circuit functionality, we allow inserting dummy edges connecting to the input of the dummy nodes but forhid using the dummy nodes to drive other nodes. Hence, we can determine *IN* and *OUT* for each node following the rule, which is listed in Fig. 2.7d. Meanwhile, because (4, 9) is the only edge connecting node 4 to unadded nodes, we have $RES_4 = \{(4, 9)\}$. When node 4 is added, all the edges in $RES_4$ will need to be lifted to BEOL.

Now, we introduce our MILP formulation for the node selection. We split the formulation into different parts to enable an easy explanation. The objective function is to minimize the cost of node selection,

$$\min_{x,d} \; \alpha \sum_i |RES_i| x_i - \beta k \sum_l (y_l + z_l) + \gamma A \sum_j d_j. \tag{2.2}$$

The cost function mainly consists of three parts: the number of edges to be lifted to BEOL layers, i.e. $\sum_i |RES_i| * x_i$, the number of edges that can be added back to the FEOL layers, i.e. $k\sum_l(y_l + z_l)$, and the area of the inserted dummy nodes $A\sum_j d_j$. $\alpha$, $\beta$, and $\gamma$ are coefficients used to control the trade-off between dummy node insertion and wire lifting. In our framework, to achieve better efficiency, a linear function is used as the optimization objective. By using a linear function, we implicitly assume that the introduced overhead is linearly dependent on the number of lifted wires and the area of the dummy nodes. Meanwhile, the cost of the lifted wires and the cost of the dummy nodes are assumed to be independent. To capture the dependency between the lift wires and the dummy nodes, a more complex nonlinear function is required, which may not be convex and can be extremely computation intensive to optimize. We empirically find that with such a linear objective function, our framework can already significantly reduce the introduced overhead compared with the existing method. We leave in-depth research on the possibility and advantage of using more complicated nonlinear functions for the optimization objective as one of our future research directions.

Now we explain the constraints. For a node $i$, it can at most be inserted into one subgraph, which is enforced by Eq. (2.3a). Meanwhile, for the $j$th subgraph, we require exactly one node to be inserted as enforced by Eq. (2.3b). We further pose the constraint in Eq. (2.3c) to ensure that exactly one node has a non-zero weight to satisfy Theorem 2.1.

$$\sum_{j=0}^{k-1} x_{ij} = x_i, \qquad \forall i; \tag{2.3a}$$

$$\sum_i x_{ij} + d_j = 1, \qquad \forall j \in \{0, \ldots, k-1\}; \tag{2.3b}$$

$$\sum_i x_i w_i = 1. \tag{2.3c}$$

Next, we need to determine the conditions for an edge to be inserted back to the FEOL layers. Consider the edge pointing from the $l$th position to the current position in $H_{s,0}, \ldots, H_{s,k-1}$. In the $j$th subgraph $H_{s,j}$, an edge pointing from $l$th position can be added back under two conditions: (1) a dummy cell is inserted to the current position; (2) node $i$ with $l \in IN_{ij}$ is inserted. Furthermore, to satisfy the requirement on subgraph isomorphism, the edge pointing from the $l$th position can be added back only when it can be added back in all the $k$ subgraphs. These two requirements can be formalized with constraints Eq. (2.4a). Note $1_{l \in IN_{ij}}$ is the indicator function that equals to 1 when $l \in IN_{ij}$ and equals to 0, otherwise. Similarly, for the edge pointing from the current position to the $l$th position, we have almost the same constraints as shown in Eq. (2.4b) except that the insertion of dummy edges pointing to the corresponding nodes is no longer allowed.

$$y_l \le y_{lj}, \quad y_{lj} \le \sum_i x_{ij} \cdot 1_{l \in IN_{ij}} + d_j, \qquad \forall j, l;$$

$$z_l \le z_{lj}, \quad z_{lj} \le \sum_i x_{ij} \cdot 1_{l \in OUT_{ij}}, \qquad \forall j, l.$$

The constraints can be further simplified by substituting $y_{lj}$ and $z_{lj}$, we have

$$y_l \le \sum_i x_{ij} \cdot 1_{l \in IN_{ij}} + d_j, \qquad \forall j, l; \tag{2.4a}$$

$$z_l \le \sum_i x_{ij} \cdot 1_{l \in OUT_{ij}}, \qquad \forall j, l. \tag{2.4b}$$

Based on the explanation above, we have the following ILP formulation for the node selection and insertion:

$$\min_{x,d} \ \text{Eq. (2.2)}$$
$$\text{s.t. Eqs. (2.3a)–(2.3c), (2.4a)–(2.4b).}$$

While all the variables in the formulation, including $x_{ij}$, $d_j$, $y_l$, and $z_l$, should be integer variables, we can relax $y_l$ and $z_l$ to be continuous without changing the optimal solution and achieve a better efficiency. By continuing the process iteratively, we can insert all the nodes with non-zero weights into the first $k$ subgraphs while keeping the $k$ subgraphs isomorphic at the same time. Then, we add all the remaining nodes into $H_{s,k}$.

### 2.5.2 Lagrangian Relaxation Algorithm

The MILP-based formulation enables us to select and insert $k$ nodes to $H_{s,0}, \ldots, H_{s,k-1}$ with a minimum cost for each iteration. However, it is still computationally expensive and suffers from unaffordable runtime for large benchmarks. We observe that two constraints that are hard to solve are Constraints (2.4a) and (2.4b). Therefore, to accelerate the framework, we apply LR to relax the last two constraints and modify the objective function as in (2.5).

$$\alpha \sum_{i,j} |RES_i| x_{ij} - \beta k \sum_l (y_l + z_l) + \gamma A \sum_j d_j$$
$$+ \sum_{j,l} \lambda_{jl} \left( - \sum_i x_{ij} \cdot 1_{l \in IN_{ij}} - d_j + y_l \right)$$

$$
\begin{aligned}
&+\sum_{j,l}\mu_{jl}\left(-\sum_{i}x_{ij}\cdot 1_{l\in OUT_{ij}}+z_l\right)\\
&=\sum_{i,j}\left(\alpha|RES_i|-\sum_{l}\lambda_{jl}\cdot 1_{l\in IN_{ij}}-\sum_{l}\mu_{jl}\cdot 1_{l\in OUT_{ij}}\right)x_{ij}\\
&+\sum_{j}\left(\gamma A-\sum_{l}\lambda_{jl}\right)d_j-\sum_{l}\left(\beta k-\sum_{j}\lambda_{jl}\right)y_l\\
&-\sum_{l}\left(\beta k-\sum_{j}\mu_{jl}\right)z_l. \qquad (2.5)
\end{aligned}
$$

Here $\mu_{jl} \geq 0$ and $\lambda_{jl} \geq 0$ are the Lagrangian multipliers. The constraints now only consist of Constraints (2.3a)–(2.3c). Compared with the original formulation, we remove the hard constraints, i.e. Constraints (2.4a) and (2.4b), and penalize the constraint violations in the objective function by updating $\lambda_{jl}$ and $\mu_{jl}$. By repeating the process of solving and updating the new formulation, the node selection algorithm will progressively converge to a legal solution to the original formulation. The proposed algorithm is summarized in Algorithm 2.

### Minimum-Cost Flow Transformation

For the new formulation, given fixed Lagrangian multipliers $\lambda_{jl}$ and $\mu_{jl}$, one important observation is that $x_{ij}$ and $d_j$ become independent with $y_l$ and $z_l$. Therefore, we can decompose the new formulation into two independent subproblems. The first subproblem is defined as below:

$$
\begin{aligned}
&\min_{x,d} \ \text{Eq. (2.5)}\\
&\text{s.t. Eqs. (2.3a)–(2.3c)},
\end{aligned}
$$

**Algorithm 2** LR-based node selection

**Input:** $k$: security level, $V$: the set of vertices to select.
**Output:** $V_{sel}$: selected vertices, $c_{sel}$: cost of vertex selection.

1: **function** `NodeSelect`($k$, $V$)
2: $\lambda_{jl} \leftarrow 0$, $\mu_{jl} \leftarrow 0$, $it \leftarrow 0$;
3: **while** $it \leq it_{max}$ **do**
4: // See Sect 2.5.2
5: $V_{sel}, c_{sel} \leftarrow$ `LagRelaxationSolve`($V, \lambda_{jl}^{it}, \mu_{jl}^{it}$);
6: // See Sect 2.5.2
7: $\lambda_{jl}^{it+1}, \mu_{jl}^{it+1} \leftarrow$ `UpdateCoeff`($\lambda_{jl}^{it}, \mu_{jl}^{it}$);
8: **end while**
9: **end function**

where $x_{ij}$, $x_i$, and $d_j$ are all binary variables. The second subproblem is defined as below:

$$\min_{x,d} \; -\sum_l \Big(\beta k - \sum_j \lambda_{jl}\Big) y_l - \sum_l \Big(\beta k - \sum_j \mu_{jl}\Big) z_l, \tag{2.6}$$

where $y_l$ and $z_l$ are the binary variables.

The solution to the second subproblem can be acquired easily as below since the objective function is monotone with $y_l$ and $z_l$, while $y_l$ and $z_l$ are independent given fixed $\lambda_{jl}$ and $\mu_{jl}$ for different $l$ in each iteration.

$$y_l = \begin{cases} 0, & \beta k - \sum_j \lambda_{jl} < 0; \\ 1, & \text{otherwise}, \end{cases}$$

$$z_l = \begin{cases} 0, & \beta k - \sum_j \mu_{jl} < 0; \\ 1, & \text{otherwise}. \end{cases}$$

For the first subproblem, one notable merit is that it can be transformed into a minimum-cost flow problem. Figure 2.8 shows an example of the constructed graph for the minimum-cost flow problem. The variables, constraints, and objectives for the first subproblem can be transformed to the concepts in the flow problem. As shown in Fig. 2.8, $V_N$ represents the set of vertices corresponding to the cells to be inserted, including the remaining nodes, i.e. nodes 4, 8, and 9, and the dummy nodes, i.e. $D_0$ and $D_1$. $V_{sub}$ denotes the set of vertices corresponding to the

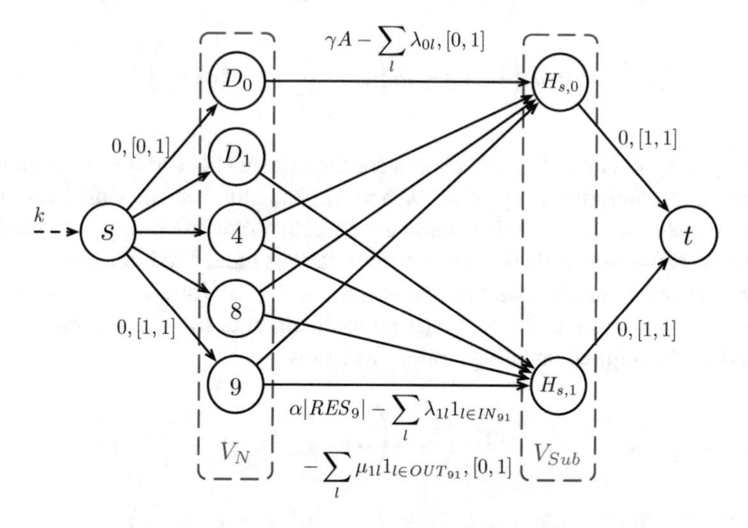


**Fig. 2.8** Example of the minimum-cost flow formulation for node selection ($k = 2$ in the example)

subgraphs, i.e. $H_{s,0}$ and $H_{s,1}$. Edges correspond to the variables in the formulation. For example, $(s, 4)$ corresponds to $x_4$, while $(4, H_{s,0})$ corresponds to $x_{40}$. Each edge is marked with the cost as well as the upper and lower bound of the capacity in Fig. 2.8. While the capacity is determined by the range of the variables, the cost for each edge is determined following the coefficients in the objective function. It should be noted that for some edges, both the upper and lower bound of the capacity is 1, which means we require a non-zero flow for the edge in the final solution. This indeed corresponds to the constants in the constraints. For example, for $(9, H_{s,1})$, which corresponds to $x_{91}$ in the original formulation, the cost becomes $\alpha|RES_9| - \sum_l \lambda_{1l} \cdot 1_{l \in IN_{91}} - \sum_l \mu_{1l} \cdot 1_{l \in OUT_{91}}$, which equals to $\alpha$ based on Fig. 2.7d. Based on the transformation above, we can easily verify all the constraints and the objective in the original formulation can be realized in the minimum-cost flow problem.

The minimum-cost flow transformation enables us to leverage efficient graph algorithms [10] to solve the originally MILP problem. As we will show in Sect. 2.6, significant runtime improvement can be achieved through the transformation.

#### Lagrangian Multiplier Update

One key step within the current node selection framework shown in Algorithm 2 is how to update the Lagrangian multiplier $\lambda_{jl}$ and $\mu_{jl}$ after each iteration. Various updating strategies may have different convergence issues. Following [14], the most widely used updating strategy for $\lambda_{jl}$ and $\mu_{jl}$ is

$$\lambda_{jl}^{it+1} = \max\left(0, \lambda_{jl}^{it} + t_{it}\left(y_l - \sum_i x_{ij} \cdot 1_{IN_{ij}} - d_j\right)\right),$$

$$\mu_{jl}^{it+1} = \max\left(0, \mu_{jl}^{it} + t_{it}\left(z_l - \sum_i x_{ij} \cdot 1_{OUT_{ij}}\right)\right),$$

where $t_{it} = 1/it^{\eta}$ is the step size chosen for the update [14] and $\eta$ is a constant.

Ideally, by iteratively updating $\lambda$ and $\mu$, the number of violations of the relaxed constraints can be reduced and the objective function in Eq. (2.5) gradually converges. However, while the number of violations indeed reduces significantly in the first several iterations, we observe severe oscillation for the objective function afterwards. To overcome the convergence problem, after the first several iterations, we modify the original updating strategy as below,

$$\lambda_{jl}^{it+1} = \lambda_{jl}^{it} + \max\left(0, t_{it}\left(y_l - \sum_i x_{ij} \cdot 1_{IN_{ij}} - d_j\right)\right),$$

$$\mu_{jl}^{it+1} = \mu_{jl}^{it} + \max\left(0, t_{it}\left(z_l - \sum_i x_{ij} \cdot 1_{OUT_{ij}}\right)\right).$$

Our updating strategy increases $\lambda$ and $\mu$ monotonically to force the value of $y_l$ and $z_l$ towards 0 in order to resolve the constrain violations and guarantee the convergence of the node selection algorithm. By controlling the maximum iteration, i.e. $it_{max}$, and the step size, i.e. $\eta$, we can control the trade-off between the solution quality and the runtime of the program.

### 2.5.3 k-Secure Layout Refinement

After solving the MILP-based formulation, cells and connections in the FEOL layers $H$ can be determined such that $\langle G, H\rangle$ is $k$-secure. The next step is to do physical synthesis to generate the layouts for the FEOL and BEOL layers. In the placement stage, existing commercial tools usually target at minimizing the total wirelength, and thus, tend to place the cells with actual connections close to each other. This, however, makes it possible for the attackers to recover the connections in the BEOL layers based on the physical proximity information [15, 24].

To guarantee the security while leveraging existing physical synthesis tools, previous method [9] chooses to ignore the lifted wires in the BEOL layers in the placement stage. For example, consider the original circuit graph shown in Fig. 2.9a, following [9], only the FEOL graph shown in Fig. 2.9b is considered in the placement stage. This helps to avoid the impact of connections in the BEOL layers, and thus, forbids the attacker from determining the identity of the nodes by physical proximity information. Though secure, this method can suffer from large overhead. This is because when the wire connections in the BEOL layers are ignored, many cells in the FEOL layers are left floating, e.g. nodes $3'$, $4'$, $D$ in Fig. 2.9b. Therefore,

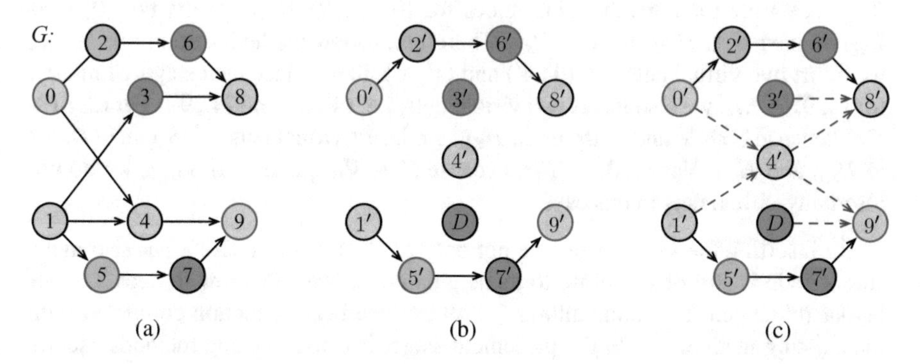


**Fig. 2.9** Comparison of the existing placement strategy and our layout refinement strategy: (**a**) original circuit graph; (**b**) existing strategy only considers the FEOL layers in the placement stage; (**c**) our strategy adds virtual nets to force a cell to be placed close to its neighbors (dotted blue lines are the virtual nets)

the distances between the cells that are actually connected in the BEOL layers, e.g. nodes $4'$ and $9'$, become highly unoptimized.

To reduce the introduced wirelength overhead, we propose a novel layout refinement technique in the placement stage. As shown in Fig. 2.9c, the basic idea of the refinement technique is to insert virtual nets between the circuit nodes that may or may not be connected in the original netlist, so that both the physical proximity between originally connected nodes and the indistinguishability among candidate nodes can be preserved.

More specifically, consider $v_i, u_j \in V_G$ with $(v_i, u_j) \in E_G$ and $i, j \in \{0, \ldots, k\}$. Their corresponding nodes $v_i', u_j'$ locate in the $i$th and $j$th subgraph of $H$, respectively, i.e. $v_i' \in V_{H_{s,i}}$, $u_j' \in V_{H_{s,j}}$. Then, depending on $i$ and $j$, there are following situations:

- When $i = j = k$, $(v_i', u_j')$ must exist in the FEOL layers, and thus, no virtual nets need to be added.
- When $i = k$ and $j \neq i$, $(v_i', u_j')$ is lifted to the BEOL layers. $\forall u_{j'}' \in V_{H_{s,i'}}$ with $u_{j'}'$ in the same position as $u_j'$ and $j' \in \{0, \ldots, k-1\}$, we insert a virtual net $(v_i', u_{j'}')$.
- When $j = k$ and $j \neq i$, $(v_i', u_j')$ is lifted to the BEOL layers. $\forall v_{i'}' \in V_{H_{s,i'}}$ with $v_{i'}'$ in the same position as $v_i'$ and $i' \in \{0, \ldots, k-1\}$, we insert a virtual net $(v_{i'}', u_j')$.
- When $i \neq k$, $j \neq k$, and $i = j$, then, $\forall v_{i'}', u_{i'}' \in V_{H_{s,i'}}$ with $v_{i'}'$ and $u_{i'}'$ in the same positions as $v_i'$ and $u_j'$, respectively, and $i' \in \{0, \ldots, k-1\}$, we insert a virtual net $(v_{i'}', u_{i'}')$.
- When $i \neq k$, $j \neq k$, and $i \neq j$, we do not insert any virtual nets.

*Example 2.5* Consider the original graph and the FEOL graph in Fig. 2.9a and c. $\{0', \ldots, 9'\}$ are the corresponding nodes for $\{0, \ldots, 9\}$, respectively. For $(0, 4) \in E_G$, we have $0' \in H_{s,0}$ and $4' \in H_{s,2}$. Therefore, following the insertion rule above, we insert two virtual nets, i.e. $(0', 4')$ and $(1', 4')$, in the placement stage. Similarly, for $(4, 9) \in E_G$, we also insert two virtual nets, i.e. $(4', 8')$ and $(4', 9')$. For $(3, 8) \in E_G$, because both $3'$ and $8'$ locate in $H_{s,0}$, we insert virtual nets $(3', 8')$ and $(D, 9')$ to $H_{s,0}$ and $H_{s,1}$. For $(1, 3) \in E_G$, because $3' \in V_{H_{s,0}}$ and $1' \in V_{H_{s,1}}$, we do not insert any virtual nets in this case.

By inserting the virtual nets, we not only guarantee the security, but also make sure a node is still placed close to its neighbors. As we will show in Sect. 2.6, our layout refinement technique allows for 49.6% overhead reduction compared with the existing method [9]. In the placement stage, because existing methods usually target at minimizing the total wirelength, cells with actual connections tend to be placed close to each other.

## 2.6 Experimental Results

### 2.6.1 Experimental Setup

In this section, we report on our experiments to demonstrate the effectiveness of the proposed split manufacturing framework. The input to our framework is a gate-level netlist and the nodes to protect. In our experiments, to select the nodes for protection, we follow the Trojan insertion methods used by TrustHub [16]. Given the netlist, we first calculate the signal probability, logic switching probability, and observability for each circuit node, and then, select the nodes with rare circuit events by comparing with a certain threshold. We modify the threshold to change the portion of nodes for protection. Our benchmarks are selected from the ISCAS benchmark suite [1] as well as the functional units (`shifter`, `alu`, and `div`) from the OpenSPARC T1 processor, the detailed statistics of which are shown in Table 2.3. In our split manufacturing scheme, following [23], FEOL layers consist of all the cells and lower metal layers up to metal 3, while BEOL layers consist of metal 4 and above. We implement our framework in C++ and use GUROBI [7] and LEMON [11] packages to solve the MILP problem and the minimum-cost flow problem, respectively. We conduct physical synthesis using Cadence Encounter [2]. All the experiments are carried out on an eight-core 3.40 GHz Linux server with 32 GB RAM. We set the runtime limit for all the algorithms to $1.5 \times 10^5$ s.

### 2.6.2 FEOL Generation Strategy Comparison

We compare the proposed MILP-based and LR-based algorithm with the previous method [9]. We set the required security level to be 10 and protect 5% of all the circuit nodes. We also set $\alpha = 0.5$, $\beta = 2.0$, and $\gamma = 0.6$. The number of LR iterations is 10 in the LR-based algorithm. We will demonstrate the impact of $\alpha$, $\beta$, and $\gamma$.

We first compare the efficiency of the three algorithms. In Table 2.3, "RT" denotes the runtime, while "$\Delta$Area" and "$\Delta$WL" denote the area and wirelength overhead compared with the original circuit. As shown in Table 2.3, on small benchmarks, compared with [9], the LR-based algorithm achieves 27,000× speedup. For large benchmarks, while [9] cannot finish within the pre-defined time threshold, our LR-based algorithm can finish within 210 s. Compared with the MILP-based algorithm, as shown in Table 2.3, the LR-based algorithm can achieve on average 9.90× speedup.

We also explore the runtime dependency of the MILP-based and LR-based algorithms on the required security level $k$ and the portion of the protected nodes. We choose the benchmark `shifter` for the study. As shown in Fig. 2.10a, b, LR-based algorithm achieves better scalability compared with the MILP-based algorithm. In Fig. 2.10b, when the portion of protected nodes exceeds 18%, the MILP-based

**Table 2.3** Runtime and overhead comparison between the MILP-based and the LR-based algorithms

| Benchmark | # PI | # PO | # Protect | # Nodes | [9] | | | MILP | | | LR-based | | |
|---|---|---|---|---|---|---|---|---|---|---|---|---|---|
| | | | | | RT (s) | $\Delta$Area (%) | $\Delta$WL (%) | RT (s) | $\Delta$Area (%) | $\Delta$WL (%) | RT (s) | $\Delta$Area (%) | $\Delta$WL (%) |
| `c432` | 36 | 7 | 11 | 214 | 414.07 | 0.0 | 175.4 | 1.029 | 1.05 | 152.4 | 0.134 | 1.03 | 165.5 |
| `c880` | 60 | 25 | 35 | 450 | 32,442 | 54.1 | 246.5 | 3.675 | 45.1 | 59.9 | 0.621 | 44.8 | 56.4 |
| `c1908` | 33 | 25 | 26 | 519 | N/A | N/A | N/A | 7.731 | 37.2 | 47.8 | 0.936 | 38.0 | 51.6 |
| `c3540` | 50 | 17 | 51 | 1012 | N/A | N/A | N/A | 40.10 | 37.9 | 96.3 | 4.080 | 37.7 | 104.5 |
| `c5315` | 178 | 101 | 94 | 1864 | N/A | N/A | N/A | 177.3 | 42.3 | 82.6 | 16.07 | 41.8 | 68.2 |
| `c6288` | 32 | 32 | 129 | 2568 | N/A | N/A | N/A | 276.1 | 38.1 | 206.9 | 32.68 | 38.4 | 220.9 |
| `shifter` | 79 | 64 | 130 | 2580 | N/A | N/A | N/A | 538.8 | 40.9 | 214.9 | 33.55 | 40.8 | 195.0 |
| `alu` | 400 | 121 | 153 | 3357 | N/A | N/A | N/A | 436.3 | 24.0 | 162.7 | 47.89 | 24.0 | 149.9 |
| `div` | 919 | 663 | 287 | 5720 | N/A | N/A | N/A | 2645.4 | 47.5 | 43.4 | 206.8 | 47.7 | 58.9 |

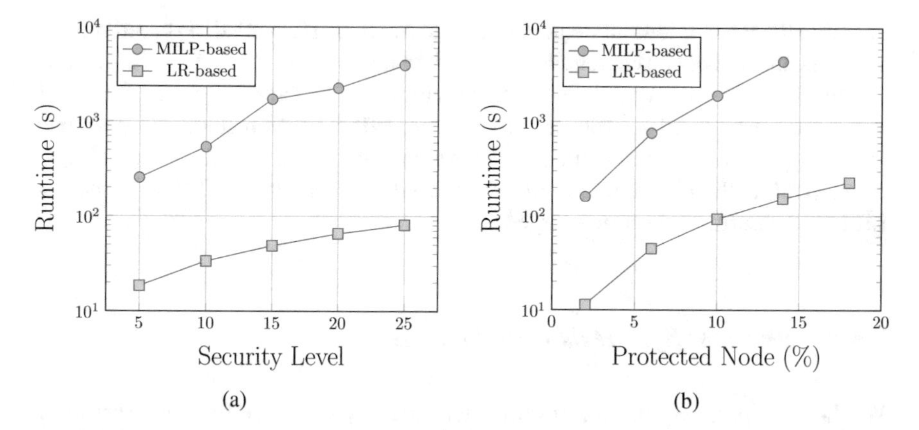


**Fig. 2.10** Runtime dependency on (**a**) the required security level and (**b**) the number of protected nodes

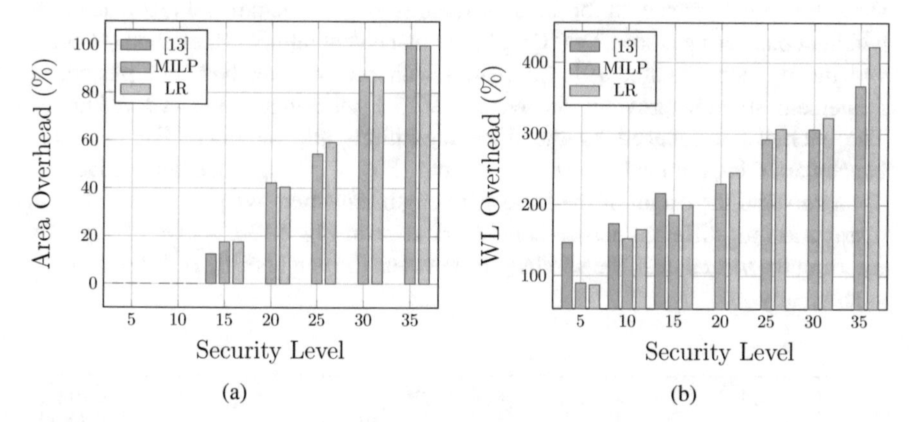


**Fig. 2.11** Compare with [9] on the (**a**) wirelength and (**b**) area overhead for different security levels

algorithm cannot be finished within the pre-defined time threshold, while it only takes 230 s for the LR-based algorithm to finish.

We then compare the overhead introduced by the three algorithms on different benchmarks as shown in Table 2.3. For the two small benchmarks, our MILP-based algorithm introduces on average 104% less wirelength overhead compared with the previous method with on average 3.97% area overhead reduction. The area and wirelength overhead introduced by the MILP-based and ILP-based algorithms are very similar.

We then compare the overhead increase with the change of the required security level $k$. We use the benchmark `c432` as an example due to the runtime limit of the previous method. As shown in Fig. 2.11, with the increase of $k$, the introduced area and wirelength overhead of all the three methods increases significantly.

Specifically, when $k$ is small, e.g., $k$ equals to 5, 10, or 15, our MILP-based method achieves much better wirelength overhead reduction with a slightly larger area overhead. When $k$ is larger than 15, the previous method cannot generate the FEOL layers for the required security level, while our MILP-based method can guarantee to achieve all the required security level. Meanwhile, we also observe that with the increase of $k$, the difference on the introduced overhead by the MILP-based and LR-based algorithms becomes larger.

### 2.6.3 Physical Synthesis Comparison

We then compare our placement refinement strategy based on the virtual net insertion with the original method proposed in [9]. The FEOL layers are generated with our MILP-based algorithm following the settings in Sect. 2.6.2. In Fig. 2.12a, we show the routed wirelength for three different strategies, including direct placement without considering $k$-security ("Orig"), our placement refinement method ("Ours") and the previous method [9]. Compared with previous method, our placement refinement strategy provides on average 97.5% wirelength overhead reduction. The overhead introduced by the three algorithms are the same. For the large benchmark `div`, our method achieves around 120% wirelength overhead reduction. To understand the origin of the large wirelength reduction, we plot the wirelength distribution for different nets in benchmark `div` in Fig. 2.12b. As we can see, by inserting the virtual nets, the wirelengths between the neighboring cells are reduced significantly.

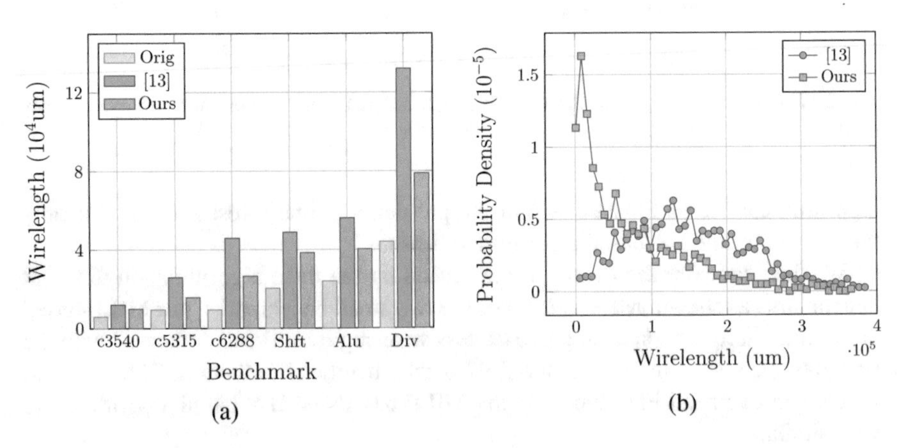


**Fig. 2.12** Overhead comparison between our layout refinement technique and [9]: (**a**) wirelength comparison, and (**b**) wirelength distribution for `div`

### 2.6.4 Physical Proximity Examination

We then carry out physical proximity checking to examine the security of the layout of the FEOL layers. In our framework, all the nodes in $H_{s,k}$ are unprotected and their identity can be determined exactly by the attackers. For example, in Fig. 2.9, node $4'$ can be identified as the corresponding node for node 4. To guarantee security, we need to prevent the attackers from identifying the protected nodes based on the identified unprotected nodes. For instance, while node $9'$ is connected to node $4'$ in the BEOL layers in Fig. 2.9, we hope that the distance between node $8'$ and $4'$ to be close to the distance between node $9'$ and $4'$. We select the benchmark `div` and set the security level to be 10. We then compare the selected nodes and their candidate nodes on the physical proximity to their neighbors. The distribution of the distance difference is shown in Fig. 2.13. As we can see, the distance difference is distributed symmetrically around 0, which indicates similar distance is achieved for the protected nodes and their candidates. This distance similarity makes the identification of the protected nodes to be nearly impossible. If we simply selected the nodes that are closest to the identified nodes, then, in all the benchmarks listed in Table 2.3, we find the number of nodes that can be correctly identified is 0. The results indicate the requirement posed by $k$-security is much higher than that of the proximity attacks, which is also the origin of the large overhead introduced to achieve $k$-security.

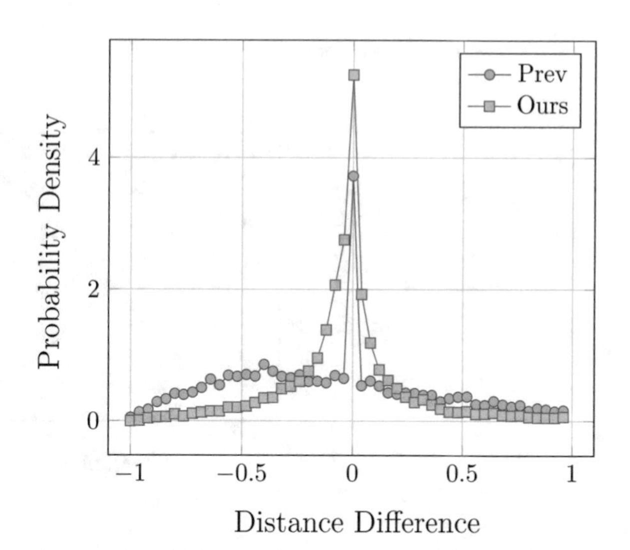


**Fig. 2.13** Distance difference comparison with [9]

### 2.6.5 Relation Between Overhead and Framework Parameters

At last, we study the change of overhead as the increase of the security level $k$, the number of protected nodes, and the coefficients $\gamma$ in the MILP formulation. We use `shifter` benchmark as an example. In Fig. 2.14a, to achieve 10-security, we show the increase of the overhead with the increase of the protected nodes. In Fig. 2.14b, we show the relation between overhead and the required security level in order to protect 5% of nodes. In Fig. 2.14c, we fix $\alpha = 0.5$, $\beta = 2.0$ in the MILP formulation and change $\gamma$ from 0.6 to 1.4. By changing $\gamma$, cell insertion and wire lifting are balanced to help provide better usage of the routing resources and the chip space for different designs.

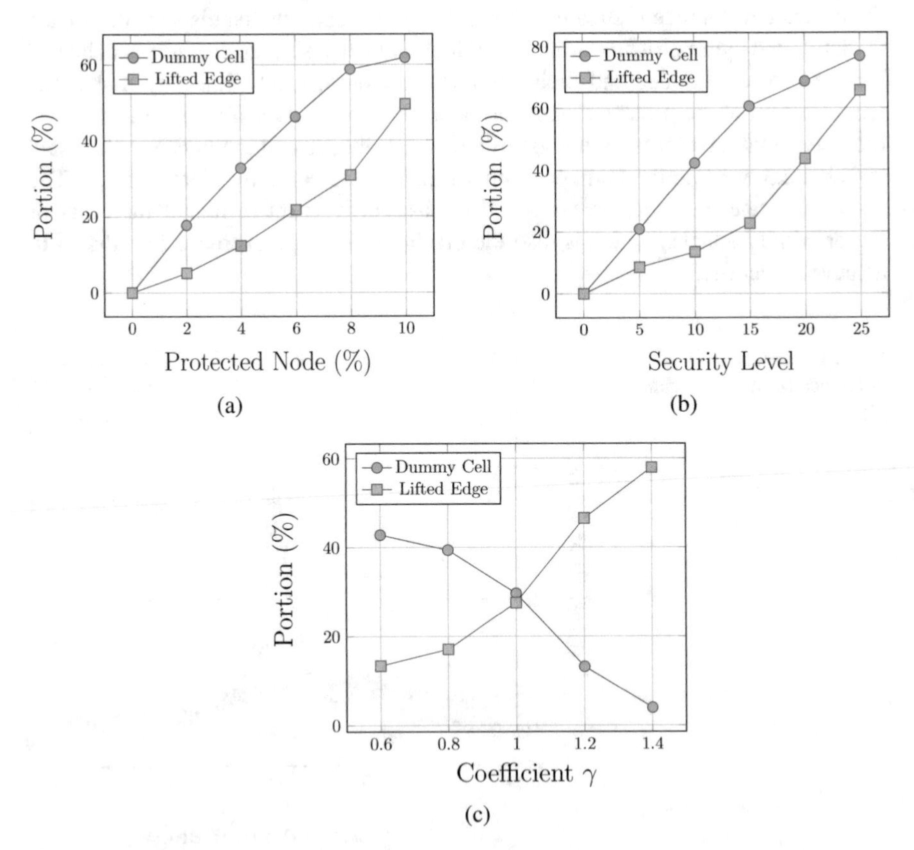


**Fig. 2.14** The relation between overhead and (**a**) portion of protected nodes, (**b**) security level, and (**c**) MILP coefficients

## 2.7 Summary

In this chapter, we propose a framework to enhance the security and practicality of split manufacturing. A new security criterion is proposed and its sufficient condition is obtained to enable more efficient realization. To realize the sufficient condition, wire lifting, dummy cell, and wire insertion are considered simultaneously through a novel MILP formulation for the first time. Layout refinement that is fully compatible with existing physical design flow is also proposed. The proposed framework achieves much better efficiency, overhead reduction, and security guarantee compared with existing methods.

## References


1. Brglez, F., Bryan, D., & Kozminski, K. (1989). Combinational profiles of sequential benchmark circuits. In *Proceedings of the IEEE International Symposium on Circuits and Systems* (pp. 1929–1934).
2. Cadence SOC Encounter. http://www.cadence.com
3. Cheng, J., Fu, A. W.-C., & Liu, J. (2010). K-isomorphism: Privacy preserving network publication against structural attacks. In *Proceedings of the ACM Conference on Management of Data*.
4. Feng, L., Wang, Y., Mak, W.-K., Rajendran, J., & Hu, J. (2017). Making split fabrication synergistically secure and manufacturable. In *Proceedings of the International Conference on Computer Aided Design*.
5. Garg,S., & Rajendran, J. J. (2017). Split manufacturing. In *Hardware protection through obfuscation* (pp. 243–262). Berlin: Springer.
6. Goldberg, A. V. (1997). An efficient implementation of a scaling minimum-cost flow algorithm. *Journal of Algorithms, 22*(1), 1–29.
7. Gurobi Optimization Inc. (2014). Gurobi optimizer reference manual. http://www.gurobi.com
8. Hill, B., Karmazin, R., Otero, C. T. O., Tse, J., & Manohar, R. (2013). A split-foundry asynchronous FPGA. In *Proceedings of the IEEE Custom Integrated Circuits Conference*.
9. Imeson, F., Emtenan, A., Garg, S., & Tripunitara, M. (2013). Securing computer hardware using 3D integrated circuit (IC) technology and split manufacturing for obfuscation. In *USENIX Security Symposium* (pp. 495–510). Berkeley: USENIX.
10. Kleinberg, J., & Tardos, E. (2005). Network flow. In *Algorithm design*. London: Pearson Education.
11. LEMON. http://lemon.cs.elte.hu/trac/lemon
12. Lin, Y., Yu, B., Xu, X., Gao, J.-R., Viswanathan, N., Liu, W.-H., et al. (2017). MrDP: Multiple-row detailed placement of heterogeneous-sized cells for advanced nodes. *IEEE Transactions on Computer-Aided Design of Integrated Circuits and Systems, 37*, 1237–1250.
13. Magaña, J., Shi, D., & Davoodi, A. (2016). Are proximity attacks a threat to the security of split manufacturing of integrated circuits? In *Proceedings of the International Conference on Computer Aided Design*.
14. Ozdal, M. M. (2009). Detailed-routing algorithms for dense pin clusters in integrated circuits. *IEEE Transactions on Computer-Aided Design of Integrated Circuits and Systems, 28*(3), 340–349.
15. Rajendran, J., Sinanoglu, O., & Karri, R. (2013). Is split manufacturing secure? In *Proceedings of the Design, Automation and Test in Europe*.

16. Salmani, H., Tehranipoor, M., & Karri, R. (2013). On design vulnerability analysis and trust benchmarks development. In *Proceedings of the IEEE International Conference on Computer Design* (pp. 471–474).
17. Sengupta, A., Patnaik, S., Knechtel, J., Ashraf, M., Garg, S., & Sinanoglu, O. (2017). Rethinking split manufacturing: An information-theoretic approach with secure layout techniques. In *Proceedings of the International Conference on Computer Aided Design*.
18. Shi, Q., Xiao, K., Forte, D., & Tehranipoor, M. M. (2017). Securing split manufactured ICs with wire lifting obfuscated built-in self-authentication. In *Proceedings of the IEEE Great Lakes Symposium on VLSI* (pp. 339–344).
19. Skorobogatov, S., & Woods, C. (2012). Breakthrough silicon scanning discovers backdoor in military chip. In *Proceedings of the International Conference on Cryptographic Hardware and Embedded Systems*.
20. Vaidyanathan, K., Das, B. P., & Pileggi, L. (2014). Detecting reliability attacks during split fabrication using test-only BEOL stack. In *Proceedings of the IEEE/ACM Design Automation Conference* (pp. 156:1–156:6).
21. Vaidyanathan, K., Liu, R., Sumbul, E., Zhu, Q., Franchetti, F., & Pileggi, L. (2014). Efficient and secure intellectual property (IP) design with split fabrication. In *Proceedings of the IEEE International Symposium on Hardware Oriented Security and Trust* (pp. 13–18).
22. Valamehr, J., Sherwood, T., Kastner, R., Marangoni-Simonsen, D., Huffmire, T., Irvine, C., et al. (2013). A 3-D split manufacturing approach to trustworthy system development. *IEEE Transactions on Computer-Aided Design of Integrated Circuits and Systems, 32*(4), 611–615.
23. Wang, Y., Cao, T., Hu, J., & Rajendran, J. (2017). Front-end of line attacks in split manufacturing. In *Proceedings of the International Conference on Computer Aided Design*.
24. Wang, Y., Chen, P., Hu, J., & Rajendran, J. (2016). The cat and mouse in split manufacturing. In *Proceedings of the IEEE/ACM Design Automation Conference*.
25. Wang, Y., Chen, P., Hu, J., & Rajendran, J. J. (2017). Routing perturbation for enhanced security in split manufacturing. In *Proceedings of the Asia and South Pacific Design Automation Conference* (pp. 605–510).
26. West, D. B. (2000). *Introduction to graph theory*. Upper Saddle River: Prentice Hall.
27. Xiao, K., Forte, D., & Tehranipoor, M. M. (2015). Efficient and secure split manufacturing via obfuscated built-in self-authentication. In *Proceedings of the IEEE International Symposium on Hardware Oriented Security and Trust* (pp. 14–19).
28. Xie, Y., Bao, C., & Srivastava, A. (2017). 3D/2.5 D IC-based obfuscation. In *Hardware protection through obfuscation* (pp. 291–314). Cham: Springer.

# Chapter 3
# IC Camouflaging Optimization and Evaluation

## 3.1 Introduction

The previous chapter has introduced a practical split manufacturing algorithm that can achieve formal security guarantee for Trojan prevention while introducing much less overhead compared with the state of the art. This chapter focuses on protecting hardware IP privacy. One of the major threats on IP privacy arises from reverse engineering [15, 22, 26, 35, 36]. As shown in Fig. 3.1, given a packaged IC acquired from the open market, after de-packaging, delayering, and image processing, the circuit layout can be reconstructed [36]. Over the last decade, the reverse engineering techniques have been developed rapidly with successful attacks demonstrated for the products of leading semiconductor companies in advanced technology nodes [5].

To thwart the IP violations and hinder the malicious attackers, IC camouflaging is proposed to hide and obscure the design details [8, 20, 27, 29, 31, 45, 50]. IC camouflaging leverages fabrication-level techniques to build camouflaging cells that can implement different Boolean logic functions, even though their layouts look the same to the attackers [8, 20, 27, 45]. By inserting the camouflaging cells into the netlist, the correct circuit functionality cannot be determined directly from the reconstructed circuit layouts.

In the past few years, the research on IC camouflaging has received considerable attention and has made remarkable progress. Existing works mainly fall into three levels: fabrication level [3, 6, 7, 37], cell level [13, 20, 21, 27], and netlist level [17, 20, 27, 32, 33, 45]. Fabrication-level camouflaging mainly focuses on developing fabrication techniques that can hide the circuit structure, including doping-based, dummy contact-based techniques, etc. Cell-level camouflaging leverages the fabrication techniques and develops different camouflaging cell designs to achieve more functionalities with indistinguishable layouts. Because the camouflaging cells usually incur overhead in terms of power, area, and timing, netlist-level strategies are proposed to develop camouflaging cell insertion algorithms to maximize the



M. Li, D. Z. Pan, *A Synergistic Framework for Hardware IP Privacy and Integrity Protection*, https://doi.org/10.1007/978-3-030-41247-0_3

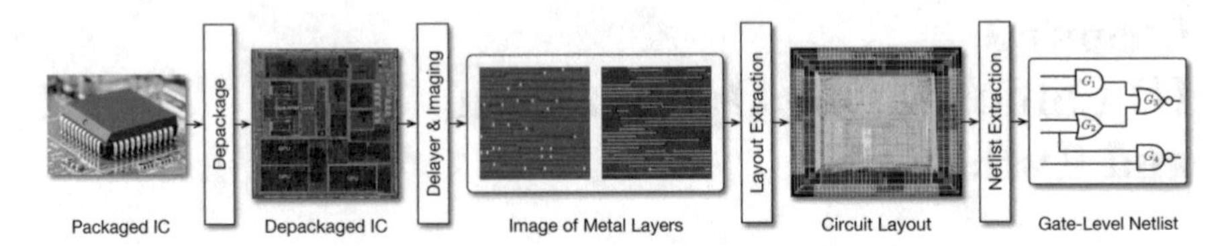


**Fig. 3.1** Physical reverse engineering flow: an adversary can de-package and delayer an IC and leverage image processing techniques to reconstruct the circuit layout. Then, the gate-level netlist can be extracted

resilience of the circuit netlist against reverse engineering attacks given the overhead constraints. For example, the authors in [17, 27, 32] insert interfered camouflaging cells or camouflaging connections to prevent reverse engineering. In [40, 43, 45], new low output-corruptibility camouflaging strategies are further proposed to protect the circuit from more advanced attack techniques [12, 34].

Simply inserting camouflaging cells does not necessarily guarantee the security against reverse engineering. Different de-camouflaging strategies, ranging from exact strategies to approximate strategies, from random strategies to active strategies, have been proposed to determine the functionality of the camouflaging cells [12, 19, 30, 34, 38, 48, 51]. In response to the fast-evolving de-camouflaging attacks, new camouflaging strategies are developed accordingly. The "arms race" inspires stronger and more rigid obfuscation algorithms and drives the evolution of the entire area.

The rest of the chapter is organized as follows. Section 3.2 defines the attack model and provides an overview of the "arms race" between the protection and attack techniques. The contribution of the chapter to the overall evolution is also discussed. Section 3.3 describes our proposed camouflaging strategy, which achieves formal security guarantee against all the exact SAT-based attack strategies. Section 3.4 illustrates our proposed attack strategy, which can efficiently reconstruct the original circuit functionality with a negligible output error probability and can be leveraged to evaluate all the existing camouflaging strategies.

## 3.2 "Arms Race" Evolution

In this section, we provide an overview of the evolution of IC camouflaging protections and attacks. The widely used attack model of IC camouflaging usually grants the attackers with the access to the following two components [12, 19, 30, 34, 38, 48, 51]:

- The obfuscated netlist, which can be acquired following the reverse engineering procedure as shown in Fig. 3.1. In the netlist, the attackers cannot determine the functionality of the camouflaging cells or the logic values of the key bits.

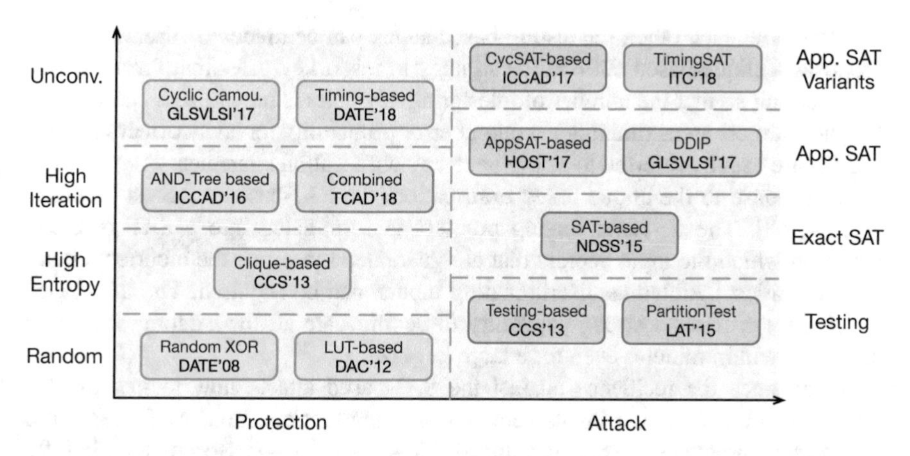


**Fig. 3.2** Overview of the "arms race" between IC camouflaging protections and reverse engineering attacks

- A functional circuit, which can be acquired from the open market and is treated as a black-box circuit. The attackers cannot observe or probe the internal signals of the functional circuit directly.

Given the camouflaged netlist, to further determine the original circuit functionality, the attackers can select a sequence of input vectors, import them into the functional circuit through the scan chain, query the functional circuit, and observe the corresponding outputs. Attackers can infer the correct circuit functionality based on the collected input–output pairs. Existing attacks are mainly different in their strategies to select the input vectors.

Over the last few years, extensive researches have been conducted for both protections and attacks. Most of the researches focus on the combinational logic with a few studies on sequential logic [11, 18, 22]. The "arms race" between the attack and protection schemes is depicted in Fig. 3.2. The first logic locking strategy, EPIC, is proposed in [29]. EPIC randomly inserts key-controlled XOR/XNOR gates into the netlist and proposes a key distribution framework using the public-key cryptography. EPIC only considers the brute force attack, which exhaustively enumerate the key space and quickly becomes intractable for a relatively large key size. Following EPIC, [2] proposes a new locking strategy that leverages key-controlled reconfigurable logic blocks, denoted as barriers. Compared with EPIC, [2] also considers brute force attacks and further develops heuristic methods to maximize the output error probability for incorrect keys.

To de-camouflage the obfuscated circuits, the testing-based attack is proposed in [27] and gets further enhanced in [17]. Instead of exhaustively searching for the whole key space, the testing-based attack leverages hardware testing principles, i.e., sensitization and justification, to select the input vectors to query the functional IC and demonstrates a strong capability to attack the randomly obfuscated circuits by isolating different key bits. However, it is observed that when different key bits can interfere with each other, the testing-based attack can be hindered. Therefore, [27] proposes a clique-based obfuscation strategy to insert key-gates that form a clique in the circuit so that the number of interfering key pairs is maximized. Because the [27] also targets at maximizing the output error probability for an incorrect key, [27] and [2] are usually referred to as high-entropy obfuscation strategies.

In response to the clique-based obfuscation, the SAT-based attack is proposed [12, 34, 48]. The de-camouflaging problem is formulated into a SAT problem, based on which the input vectors that are guaranteed to prune the incorrect circuit functionalities, denoted as discriminating inputs, can be acquired. The SAT-based attack demonstrates a strong capability to deobfuscate all the existing protection strategies within minutes even for a large key size.

To enhance the resilience against the SAT-based attack, how to increase the number of discriminating inputs becomes an important question. AND-tree based obfuscation strategies are thus developed [20, 40, 44, 45, 47]. Specifically, in [20], for the first time, the equivalence between the deobfuscation attack and the Boolean function learning problem is built. Based on the equivalence, it is formally proved that AND-tree based strategies require an exponential number of discriminating inputs to deobfuscate.

Although the AND-tree based strategies achieve an exponential increase of resilience against the SAT-based attacks with respect to the AND-tree size, [30] observes that the output error probability for an incorrect key also reduces exponentially. Hence, an approximate SAT-based attack, denoted as the AppSAT attack, is proposed, which can efficiently obtain a netlist that functions correctly with very high probability. The AppSAT attack forces the protection to consider both the output error probability and the number of discriminating inputs.

To defend against the AppSAT attack, the combined strategies that leverage both the clique-based and AND-tree based methods are proposed, which, however, show limited resilience against the AppSAT attack. Meanwhile, [32, 33, 50] propose to introduce unconventional structures for circuit obfuscation. In [28, 32, 33], circuit wire interconnections are camouflaged by introducing dense, nested, "fake" cyclic structures into the netlist, which does not impact the circuit functionality but significantly complicates the AppSAT attack process. Zhang et al. [50] proposes to deliberately remove the flip flops in the circuit to create a mixture of single-cycle and multiple-cycle paths in the circuit, which cannot be directly resolved by the AppSAT attacks either. Though the corresponding enhancements for the AppSAT attack are proposed, i.e., CycSAT [51] and TimingSAT [19], the attack efficiency is significantly degraded due to the unconventional circuit structures.

The description above illustrates the evolution of logic obfuscation protections and attacks. The "arms race" inspires more rigid and secure protection schemes and significantly boosts the advancement of the whole area. The dissertation makes specific contributions to both camouflaging protections and reverse engineering attacks [19, 20, 30, 32, 33]. Specifically, in Sects. 3.3 and 3.4, the AND-tree based camouflaging strategy proposed in [20] and the TimingSAT proposed in [19] will be described in detail.

## 3.3 Provably Secure IC Camouflaging

As described in Sect. 3.2, traditional high-entropy camouflaging strategies [2, 27] have fundamental problems. First, due to the lack of provably secure criteria to guide IC camouflaging, high-entropy strategies usually tend to over-estimate the provided security level and, in fact, have been shown vulnerable to the existing SAT-based de-camouflaging attacks as well as removal attacks based on structural and functional information [12, 34, 46, 48]. Second, the insertion of camouflaging cells usually leads to large overhead, which places significant limits on their usage in commercial applications.

In this section, a new criterion, defined as de-camouflaging complexity, to directly quantify the security of the camouflaged netlist is proposed. The proposed security criterion is defined as the number of input–output patterns that an attacker has to evaluate to decide the functionality of the original circuit and is formally analyzed by showing the equivalence between SAT-based de-camouflaging attack and the active learning scheme [9, 10, 14]. The proposed security criterion is favorable as it is independent of how the SAT-based attack is formulated or what type of machine is used for the attack. The equivalence also enables us to identify two key factors that determine the security of a camouflaged netlist.

To increase the security level of the camouflaged netlist, we propose two camouflaging strategies targeting at the two identified factors. The first camouflaging strategy is a new low-overhead camouflaging cell generation, which is mainly based on the observation that the overhead of a camouflaging cell depends on its actual functionality in the circuit. We create a specific kind of camouflaging cell that incurs negligible overhead for one functionality, which allows for a large amount of insertion into the netlist and thus provides better security protection. The second camouflaging strategy leverages the AND-tree structure for better security. We analyze the stand-alone AND-tree structure to verify its induced exponential increase of security level and further identify two important properties, denoted as tree decomposability and input bias, both of which are important to guarantee its effectiveness in general circuits. Combining these two strategies together, an IC camouflaging framework is then proposed to further optimize the camouflaged circuit for better protection against removal attacks. Experimental results demonstrate that the functionality of the camouflaged netlist generated by our framework cannot be resolved by existing de-camouflaging techniques and the overhead is negligible. The contributions are summarized as follows:

- We investigate a new security criterion to quantify the de-camouflaging complexity and identify two key factors that can help enforce the security criterion in camouflaged netlist.
- We propose two novel camouflaging strategies to increase the two identified factors.
- We develop an IC camouflaging framework combining the two strategies to further protect the camouflaged circuits against removal attacks.

- We verify our proposed security criterion and framework against state-of-the-art de-camouflaging techniques and demonstrate great resilience with negligible overhead.

The rest of Sect. 3.3 is organized as follows. Section 3.3.1 provides a review on the preliminaries on active learning scheme. Section 3.3.2 formally builds the equivalence between SAT-based de-camouflaging and active learning with key security factors identified. Sections 3.3.3 and 3.3.4 describe the camouflaged cell generation strategy and the AND-tree structure. Section 3.3.5 proposes an IC camouflaging framework. Section 3.3.6 demonstrates the performance of the proposed camouflaging framework, followed by conclusion in Sect. 3.3.7.

### *3.3.1 Preliminary: Active Learning*

In this section, we provide basic definitions concerning active learning. For more detailed description, interested readers can refer to [9].

Considering an arbitrary domain $X$ where a concept $h$ is defined to be a subset of points in the domain, a point $x \in X$ can be classified by its membership in concept $h$, that is, $h(x) = 1$ if $x \in h$, and $h(x) = 0$ otherwise. A concept class $H$ is a set of concepts. For a target concept $t \in H$, a training sample is a pair $(x, t(x))$ consisting of a point $x$, which is drawn from $X$ following distribution $\mathscr{D}$, and its classification $t(x)$. A concept $h$ is defined to be consistent with a sample $(x, t(x))$ if $h(x) = t(x)$.

The intuition of active learning is to regard learning as a sequential process, so as to choose samples adaptively. Consider a set $S$ of $m$ samples. The classification of some regions of the domain can be determined, which means all concepts in $H$ that are consistent with $S$ will produce the same classification for the points in these regions. Active learning scheme seeks to avoid sampling new points from these regions, and instead, samples only from the regions that contain points which can have different classifications for different concepts in $H$, denoted as region of uncertainty $\mathscr{R}(S)$. By iteratively sampling from $\mathscr{R}(S)$ and updating $\mathscr{R}(S)$ based on the new sample, $t$ can be learned from $H$. We use the following example to illustrate the concept of active learning.

*Example 3.1* Consider a two-dimensional space, and the target $t$ is a set of points lying inside a fixed rectangular in the plane as shown in Fig. 3.3. Assuming we already have some samples with their classification, $\mathscr{R}(S)$ can then be decided accordingly. Consider the three points $s_1$, $s_2$, and $s_3$ in Fig. 3.3, the label for $s_1$ and $s_2$ can already be determined based on existing samples. Therefore, only $s_3$ can help provide further information to decide the target $t$ from the concept class $H$.

According to [9], if we define error rate $\mathrm{er}_{x\sim\mathscr{D}}(h, t)$ for a concept $h$ with respect to the target $t$ and the distribution $\mathscr{D}$ of points $x$ as $\mathrm{er}_{x\sim\mathscr{D}}(h, t) = \mathsf{Pr}_{x\sim\mathscr{D}}[h(x) \neq t(x)]$, then by adaptively sampling from $x \in X$, to guarantee $\mathrm{er}_{x\sim\mathscr{D}}(h, t) \leq \epsilon$ with sufficient probability, the number of samples $m$ needed for active learning is

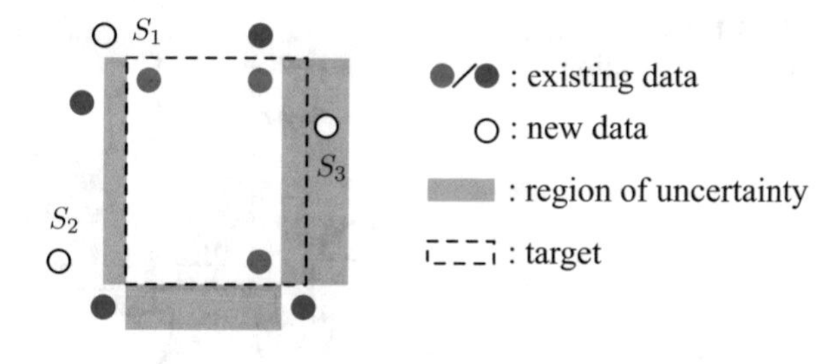


**Fig. 3.3** Example of sampling strategy for active learning

$$m = \mathcal{O}\left(\theta d \log\left(\frac{1}{\epsilon}\right)\right),$$

where $d$ is a measure of the capacity of $H$. Specially, when $X$ is a Boolean domain with $X = \{0, 1\}^n$ and the concept class contains only Boolean function, we have $d \geq \frac{\log_2|H|}{n}$ [39]. Here $|\cdot|$ denotes the cardinality of the set. $\theta$ is the disagreement coefficient, defined as

$$\theta = \sup_{\epsilon} \frac{\mathsf{Pr}_{x\sim\mathscr{D}}[\mathrm{DIS}(H_\epsilon)]}{\epsilon},$$

where $H_\epsilon = \{h \in H : \mathrm{er}_{x\sim\mathscr{D}}(h, t) \leq \epsilon\}$, $\mathrm{DIS}(H_\epsilon) = \{x : \exists h, h' \in H_\epsilon \text{ s.t. } h(x) \neq h'(x)\}$, and $\mathsf{Pr}_{x\sim\mathscr{D}}[\mathrm{DIS}(H_\epsilon)] = \mathsf{Pr}_{x\sim\mathscr{D}}[x \in \mathrm{DIS}(H_\epsilon)]$.

### *3.3.2 IC Camouflaging Security Analysis*

Let $c_o$ be the original circuit before camouflaging. $c_o$ has $n$ input bits with the input space $I \subseteq \{0, 1\}^n$ and $l$ output bits with output space $O \subseteq \{0, 1\}^l$. Define the indicator function $e_{c_o} : I \times O \to \{0, 1\}$ for $c_o$, where $I \times O = \{(i, o) : i \in I, o \in O\}$, as

$$e_{c_o}(i, o) = \begin{cases} 1, & \text{if } c_o(i) = o, \\ 0, & \text{otherwise,} \end{cases}$$

where $e_{c_o}$ indicates whether an output vector $o$ can be generated by $c_o$ given certain input vector $i$.

During the process of IC camouflaging, $\widetilde{m}$ camouflaging gates are inserted into the original netlist, whose functionalities cannot be resolved by physical reverse engineering techniques. Let $G$ denote the set of all possible functionalities for the camouflaging gate, where $\forall g \in G$, $g : \{0, 1\}^{\widetilde{n}} \to \{0, 1\}$ with $\widetilde{n}$ as the input number of the camouflaging gate. Let $y$ denote $\widetilde{m}$ functions chosen from $G$, i.e. $y \in G^{\widetilde{m}}$, which assigns each camouflaging gate a function in $G$ and let $Y$ denote the set of all

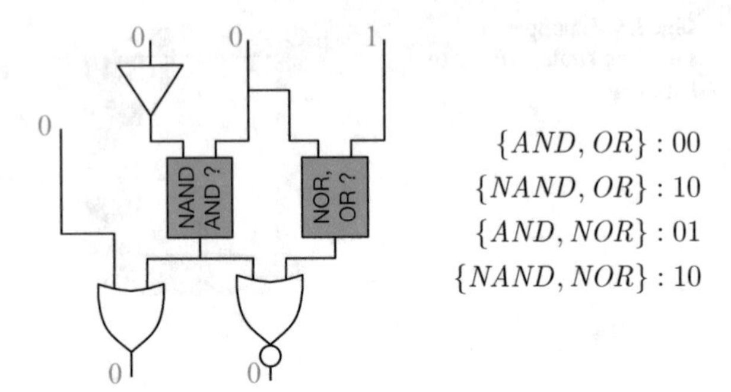


**Fig. 3.4** Example of the camouflaged netlist

possible $y$. Depending on $y$, a set of possible circuit functionalities can be created, denoted as $C$. Note that $c_o \in C$. $\forall c \in C$, there exists a corresponding indicator function $e_c$. Let $E_C$ denote the set of indicator functions for all $c \in C$.

Based on the attack model described in Sect. 3.3.1, after physical reverse engineering, the attacker can acquire the camouflaged netlist but cannot resolve the functionality of the camouflaging cells. Equivalently, the attackers can acquire $C$ and $E_C$ from physical reverse engineering. For the attackers, to resolve $c_o \in C$ is equivalent to resolving $e_{c_o} \in E_C$. The attacker can select input pattern $i \in I$, apply to the black-box functional circuit through circuit scan chain, and get the correct output $c_o(i)$. Based on $(i, c_o(i))$, all $c \in C$ that are not consistent with $(i, c_o(i))$ can be pruned.

*Example 3.2* Consider the camouflaged circuit shown in Fig. 3.4. We have $Y = \{\{\text{NAND}, \text{NOR}\}, \{\text{AND}, \text{NOR}\}, \{\text{NAND}, \text{OR}\}, \{\text{AND}, \text{OR}\}\}$. Assume {AND, OR} gives the correct circuit functionality, then, for input pattern {0001}, by evaluating the black-box functional circuit, the correct output vector should be {00}. This indicates $e_{c_o}(0001, 00) = 1$. For circuit $c_y$ with $y = \{\text{NAND}, \text{OR}\}$, given input pattern {0001}, the output becomes {10}; Therefore, $e_{c_y}(0001, 10) = 0$. The indicator functions of the other two functionalities both equal to 0 given ({0001}, {00}). Therefore, the input–output pattern ({0001}, {00}) can help rule out all the false functionalities of the camouflaged circuits.

To evaluate the effectiveness of camouflaging and the hardness of de-camouflaging, we define the de-camouflaging complexity as the number of input–output patterns required to rule out the false functionalities and resolve $c_o \in C$, equivalently $e_{c_o} \in E_C$. To evaluate the de-camouflaging complexity, we build the equivalence between the SAT-based de-camouflaging strategy and the active learning scheme as follows:

- The set of indicator functions of all possible circuit functionalities $E_C$ corresponds to the concept class $H$;
- The supply of indicator functions, i.e. $I \times O$, corresponds to the set of points $X$;
- The indicator function of the original circuit functionality $e_{c_o}$ corresponds to the target concept $t$;

- The input–output relation $((i, c(i)), 1)$ corresponds to the samples $(x, t(x))$;
- The SAT-based de-camouflaging strategy corresponds to the selective sampling strategy.

Based on the equivalence, the number of input–output patterns required to resolve $e_{c_o}$ with less than $\epsilon$ error rate and sufficiently high probability is

$$m(e_{c_o}, E_C) = \mathcal{O}\left(\theta d \log\left(\frac{1}{\epsilon}\right)\right), \tag{3.1}$$

where $d \geq \frac{\log_2|E_C|}{n}$ is related to the number of functionalities in $E_C$. $\theta$ is calculated as

$$\theta = \sup_\epsilon \frac{\Pr_{(i,o)\sim I\times O}[(i, o) \in \mathrm{DIS}(E_\epsilon)]}{\epsilon}, \tag{3.2}$$

where $E_\epsilon = \{e_c \in E_C : \mathrm{er}_{(i,o)\sim I\times O}(e_c, e_{c_o}) \leq \epsilon\}$ consists of all the indicator functions that are different from $e_{c_o}$ with probability less than $\epsilon$, and $\mathrm{DIS}(E_\epsilon) = \{(i, o) : \exists e_c, e_{c'} \in E_\epsilon \text{ s.t. } e_c(i, o) \neq e_{c'}(i, o)\}$ consists of all the input–output pairs $(i, o)$ that lead to different outputs of any pair of indicator functions in $E_\epsilon$. We use the following example to illustrate the $E_\epsilon$ and $\mathrm{DIS}(E_\epsilon)$.

*Example 3.3* Consider the camouflaged circuit and the truth table of all the possible functionalities of the camouflaged circuit shown in Fig. 3.5. The correct functionality is $c_0$ with $y^* = \{\mathsf{BUF}, \mathsf{BUF}\}$. Then, for $c_0$, the indicator function $e_{c_0}$ becomes

$$e_{c_0}(i, o) = \begin{cases} 1, & \text{if } (i, o) \in \{(00, 0), (01, 0), (10, 0), (11, 1)\}, \\ 0, & \text{otherwise.} \end{cases}$$

Similarly, we can define $e_{c_i}$ for $c_i$, $1 \leq i \leq 3$. $e_{c_1}$ has different outputs compared with $e_{c_0}$ at four input–output pairs, i.e., $\{(10, 0), (10, 1), (11, 0), (11, 1)\}$. If we assume $(i, o)$ follows a uniform distribution, then $\mathrm{er}_{(i,o)\sim I\times O}(e_{c_0}, e_{c_1}) = 4/8 = 1/2$. Similarly, we have $\mathrm{er}_{(i,o)\sim I\times O}(e_{c_0}, e_{c_2}) = \mathrm{er}_{(i,o)\sim I\times O}(e_{c_0}, e_{c_2}) = 1/2$. If we set $\epsilon = 1/2$, $E_{1/2} = \{e_{c_0}, e_{c_1}, e_{c_2}, e_{c_3}\}$. We can also determine $\mathrm{DIS}(E_{1/2}) = \{(00, 0), (00, 1), (01, 0), (01, 1), (10, 0), (10, 1), (11, 1), (11, 1)\}$

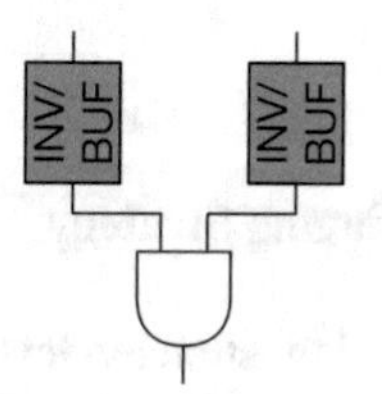


| Input | $c_0$ | $c_1$ | $c_2$ | $c_3$ |
|---|---|---|---|---|
| 00 | 0 | 0 | 0 | 1 |
| 01 | 0 | 0 | 1 | 0 |
| 10 | 0 | 1 | 0 | 0 |
| 11 | 1 | 0 | 0 | 0 |

**Fig. 3.5** Example of a camouflaged netlist and the truth table for all the possible functionalities

and $\Pr_{(i,o)\sim I\times O}[(i,o) \in \mathrm{DIS}(E_{1/2})] = 1$. By trying different $\epsilon$, we know $\theta$ get the maximum value, i.e., 2, when $\epsilon = 1/2$.

Based on the equivalence of SAT-based attack and active learning, we can identify two key factors that impact the security of different camouflaging strategies, i.e. $d$ and $\theta$, and also use Eq. (3.1) as a quantitative security evaluation metric. It should be noted that Eq. (3.1) refers to a specific scenario defined as probably approximately correct (PAC) learning, which indicates the output of active learning scheme is an approximation of the original functionality with certain probability. However, SAT-based attack is an exact learning scheme. Though different, the exact learning problem is at least as difficult as the PAC learning problem, which indicates Eq. (3.1) can still work as a lower bound of the complexity of the SAT-based attack. In fact, exact learning may not always be necessary. From the attacker's perspective, it is sufficient if he can resolve a circuit functionality, whose output error probability is limited to certain small threshold compared with original circuit functionality. We regard this as a future research direction. In the following sections, we will propose camouflaging techniques to increase $d$ and $\theta$ to enhance the resilience against SAT-based attack.

### 3.3.3 *Novel Camouflaging Cell Design*

In this section, we target at increasing $d$ as in Eq. (3.1). Because the lower bound of $d$ is in proportional to $|C|$, i.e. $|E_C|$, we choose to increase the number of possible functionalities of the camouflaged netlist. $|E_C|$ is related to the number of camouflaging cells inserted into the circuits, the locations of inserted cells, and the number of possible functionalities of different camouflaging cells. Traditional camouflaging cell generation strategies usually target at increasing the possible functionalities for each cell. However, they usually cause large overheads in terms of power, area, and timing, which significantly limits the number of camouflaging cell that can be inserted into the original netlists, and, thus, limits the total number of possible functionalities of the camouflaged netlists. We observe that *the overhead of one camouflaging cell is mainly determined by its actual functionality in the circuit.* In this section, we will propose two different camouflaging cell designs, termed as XOR-type cells and stuck-at-fault-type (STF-type) cells, which incur negligible overhead for some specific functionalities. As we will show, the proposed designs allow much richer functionalities for the camouflaged netlist with negligible overhead.

#### XOR-Type Cell Camouflaging Strategy

The XOR-type camouflaging strategy leverages the dummy contact technique. For example, as shown in Fig. 3.6a, for a BUF cell, we modify the shape of the

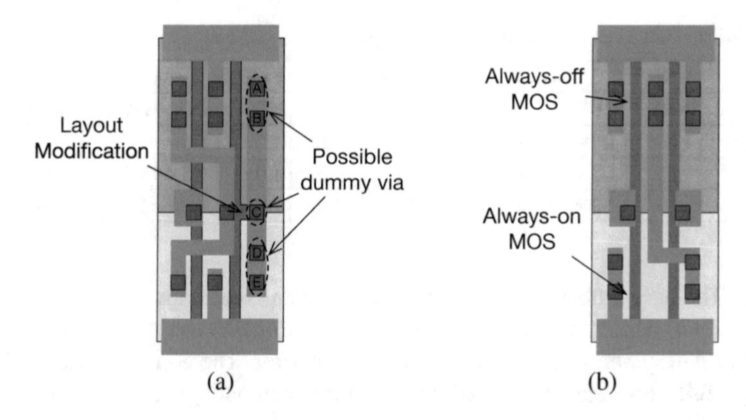


**Fig. 3.6** Examples of two different cell camouflaging strategies: (**a**) XOR-type and (**b**) STF-type

**Table 3.1** Overhead characterization of XOR-type camouflaged cell

| Cell | BUF | | AND2 | | OR2 | | AND3 | | OR3 | |
|---|---|---|---|---|---|---|---|---|---|---|
| Act. func. | BUF | INV | AND2 | NAND2 | OR2 | NOR2 | AND3 | NAND3 | OR3 | NOR3 |
| Timing | 1.0× | 2.0× | 1.0× | 1.5× | 1.0× | 1.9× | 1.0× | 1.8× | 1.0× | 1.9× |
| Area | 1.0× | 1.5× | 1.0× | 1.3× | 1.0× | 1.3× | 1.0× | 1.3× | 1.0× | 1.3× |
| Power | 1.0× | 1.5× | 1.0× | 0.9× | 1.0× | 1.1× | 1.0× | 1.0× | 1.0× | 1.0× |

polysilicon to create extra overlap between polysilicon and metal layer. Then, we can configure the functionality of the camouflaging cell by determining whether the five contacts are real or dummy. When contacts $A$, $B$, $D$, $E$ are real, while contact $C$ is dummy, the cell functions as a BUF. If contact $C$ is real, while the rest of the contacts are dummy, the cell functions as an INV. Similar strategy can be applied to other cells including AND, OR, and so on.

To evaluate the overhead of the XOR-type camouflaged cells, standard cells from NanGate 45 nm Open Cell Library [25] are modified according to the strategy and scaled to 16 nm technology. Then Calibre xRC [23] is used to extract parasitic information of the cell layouts. We use SPICE simulation to characterize different types of gates, which are based on 16 nm PTM model [1]. As we can see from Table 3.1, when the cell functions as a BUF, the overhead induced by the layout modification is negligible compared with original standard cells. However, when the cell functions as an INV, large overhead can be observed for timing, area, and power.

#### STF-Type Cell Camouflaging Strategy

The STF-type camouflaging strategy leverages the doping-based camouflaging technique. The camouflaging cell generated with the STF-type strategy has exactly the same metal and polysilicon layer compared with the existing standard cells in the

**Table 3.2** Overhead characterization of STF-type camouflaged cell

| Cell | AND2 | | OR2 | | NAND2 | | NOR2 | | AND3 | | |
|---|---|---|---|---|---|---|---|---|---|---|---|
| Act. func. | AND2 | BUF | OR2 | BUF | NAND2 | INV | NOR2 | INV | AND3 | AND2 | BUF |
| Timing | 1.0× | 1.4× | 1.0× | 1.4× | 1.0× | 1.6× | 1.0× | 1.6× | 1.0× | 1.3× | 1.8× |
| Area | 1.0× | 1.3× | 1.0× | 1.3× | 1.0× | 1.5× | 1.0× | 1.5× | 1.0× | 1.3× | 1.7× |
| Power | 1.0× | 1.2× | 1.0× | 1.2× | 1.0× | 1.5× | 1.0× | 1.5× | 1.0× | 1.1× | 1.3× |

library. The only difference comes from the type and the shape of the lightly doped drain (LDD), which makes it very difficult to distinguish a regular MOS transistor with the Always-on and Always-off MOS transistor. The STF-type camouflaging strategy fully leverages this flexibility to create camouflaging cells with different functionalities.

For example, as shown in Fig. 3.6b, for a NAND2 cell, if we change the doping scheme following [3], we can create Always-on NMOS transistor and Always-off PMOS transistor associated with $A$. This is equivalent to creating a stuck-at-1 fault at input $A$ and the functionality of the NAND cell becomes an INV for input $B$. Similar strategy can be applied to all the other cells in the original library. In terms of overhead, it is obvious that the STF-type camouflaging strategy does not impact the timing or performance when it functions normally since the layout is not modified. However, when Always-on or Always-off scheme is used, the overhead needs to be characterized. We use the same method as described for the XOR-type camouflaging strategy and the overhead results are listed in Table 3.2.

## Discussion

As described above, both XOR-type and STF-type camouflaging cells proposed above incur negligible overhead for some specific functionalities. It should be noted that they also have different characteristics. For the XOR-type camouflaging cell, when the attacker mis-interprets the type of the contact, the probability of logic error at the output of the cell is always 1. For the STF-type cell, a mis-interpretation of the doping scheme may not always lead to incorrect logic value at the output of the gate. For example, consider an AND gate with $\widetilde{n}$ inputs, denoted as $i_1, i_2, \ldots, i_{\widetilde{n}}$, and first $\widetilde{n}'$ inputs are dummy. Then, the probability of logic error at the output of the cell can be calculated as

$$P_e = \Pr_{i \sim I}\left[\left(\bigcup_{k \in [\widetilde{n}']} i_k = 0\right) \cap \left(\bigcap_{k \in [\widetilde{n}] \backslash [\widetilde{n}']} i_k = 1\right)\right],$$

where $[\widetilde{n}] = \{1, 2, \ldots, \widetilde{n}\}$.

Meanwhile, for the STF-type camouflaging cell, since it is equivalent to creating a stuck-at fault at several input pins of the cell, these input pins become dummy as their logic values do not impact the output of the cell. This enables us to create dummy wire connections between different nodes that are not connected in the

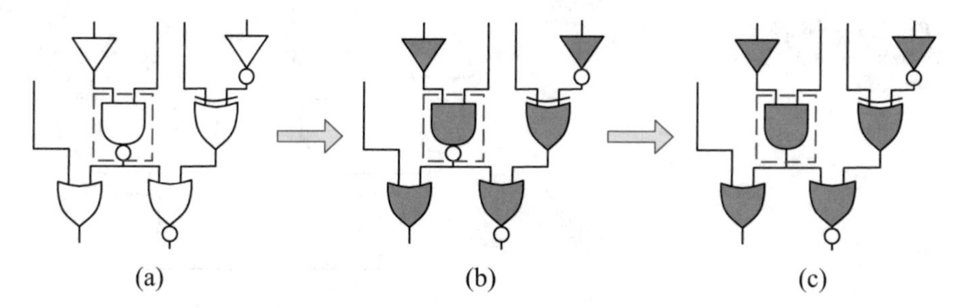


**Fig. 3.7** Two-step IC camouflaging with XOR-type and STF-type cells: (**a**) original circuit netlist; (**b**) change all standard cells into camouflaging cells that appear to be same and have the same functionalities; (**c**) random select cells and replace them with cells that appear to be different but work with the same functionalities

original circuit, which not only hide the original functionality, but also hide the circuit structure.

To leverage the XOR-type and STF-type cells to camouflage the original circuits, we propose a two-step strategy. In the first step, we replace all the standard cells with the camouflaging cells, e.g. NAND cell to an STF-type NAND cell in Fig. 3.7b. For these camouflaging cells, they are set to work as the cells that they appear to be, e.g. an STF-type NAND cell works as a real NAND gate, and, therefore, the replacement incurs negligible overhead based on our characterization results above. Then, in the second step, we randomly choose a small subset of gates in the netlist, and replace them with new camouflaging cells that appear differently but indeed work with the same functionality as the original cells, e.g. a NAND cell is replaced by an XOR-type AND cell in Fig. 3.7c. Although overhead can be introduced in the second step, we argue such overhead can be negligible since only a small subset of gates are modified in the second step. Even if we assume the attackers know the number of cells that are changed in the second step, they cannot determine which cells are changed. Therefore, the attacker still cannot determine the functionality for each cell in the camouflaged circuits. With the increase of circuit size, the total number of possible functionalities of the camouflaged netlist also increases, and, thus, results in high resilience towards SAT-based attack as shown in our experimental results.

The effectiveness of the proposed camouflaging cell generation strategy is verified in Sect. 3.3.6. However, there are several drawbacks if we simply use the cell generation strategy to protect the circuits:

- Evaluating $|C|$ or $|E_C|$ accurately is usually computational intractable, which makes it hard to provide provably secure guarantee.
- Since we only leverage camouflaging cells to replace cells in the original circuit, the total number of camouflaging cells is limited by the circuit size. Empirically, for large circuit, the proposed strategy works well, but for small circuits the correct circuit functionality can still be resolved.
- Even for large circuit, it is hard to protect all the circuit outputs, which means it is possible for the attacker to slice the circuits for different primary outputs,

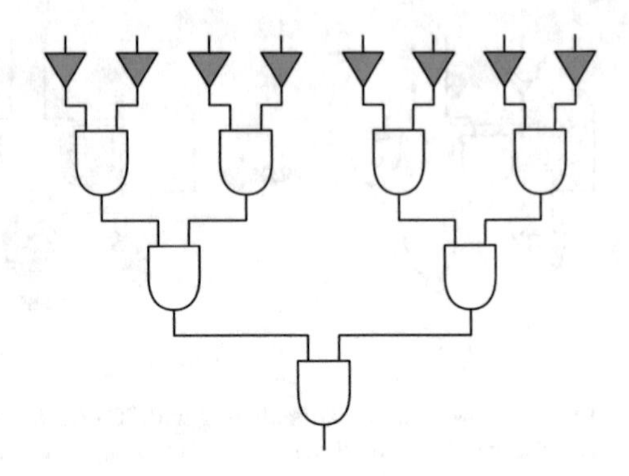

**Fig. 3.8** Example of a camouflaged AND-tree structure

and individually de-camouflage the functionality of some circuit primary outputs following [17].

To overcome these problems, in Sect. 3.3.4, we will propose another camouflaging technique based on AND-tree structure, which provides provably secure guarantee.

### 3.3.4 AND-Tree Camouflaging Strategy

In this section, we target at increasing $\theta$ as in Eq. (3.1). In [34], the AND-tree structure is noticed to achieve good resilience to SAT-based de-camouflaging attack when the input pins are camouflaged as shown in Fig. 3.8. In this section, we provide formal analysis for the AND-tree structure and further identify two important characteristics of the AND-tree structure, denoted as input bias and tree decomposability, to characterize its effectiveness in general circuits.

#### Security Analysis of the AND-Tree Structure

Consider the AND-tree structure with $n$ input pins shown in Fig. 3.8, where all the input pins are camouflaged with the XOR-type camouflaging BUF cell. Recall from Sect. 3.3.2 that $I \subseteq \{0, 1\}^n$ and $Y \subseteq G^n$ represent all the possible combinations of functionalities for the camouflaging cells. For any $i \in I$ and $y \in Y$, the output of the AND-tree structure can be expressed as

$$c_y(i) = g_1(i_1) \wedge g_2(i_2) \wedge \ldots \wedge g_n(i_n),$$

where $i_k$ denotes the $k$th entry of input $i$, and $g_k(\cdot)$ denotes functionality of the $k$th camouflaging cell. $g_k(i_k) = i_k$ if the $k$th cell functions as a buffer, while $g_k(i_k) = \overline{i_k}$ if the $k$th cell functions as an inverter.

Let $y^* \in Y$ denote the correct configuration for all the camouflaging cells. Then, depending on $y$, there are $2^n$ different circuit functionalities, i.e. $|C| = |E_C| = 2^n$. For any $y \in Y$, there exists exactly one input $i \in I$ such that $c_y(i) = 1$, denoted as $i^y$. Therefore, we have $\Pr_{i\sim I}[c_y(i) = 1] = \Pr_{i\sim I}[i = i^y]$. Now, we have the following lemma for the camouflaged AND-tree structure.

**Lemma 3.1** *For an n-bit AND-tree structure with all tree inputs camouflaged with XOR-type camouflaging BUF cells, if the logic values for tree inputs follow identical independent Bernoulli distribution with probability of* 0.5*, then, we have* $\theta = 2^{n-1}$.

To prove Lemma 3.1, we will first demonstrate that when the logic value for all the tree inputs follow identical independent Bernoulli distribution with probability of 0.5, for any $y \neq y^*$, the error rate of the indicator function $e_{c_y}$ is $1/2^{n-1}$ compared with $e_{c_{y^*}}$. Meanwhile, we will show that $\text{DIS}(E_\epsilon) = I \times O$. Then, based on the definition of $\theta$ in Eq. (3.2), we will prove that $\theta = 2^{n-1}$.

***Proof 3.1*** For any $y \neq y^*$, $c_y$ is different compared with $c_{y^*}$ for exactly two input vectors, i.e. $i^y$ and $i^{y^*}$. For $i^y$, because $c_y(i^y) = 1$ while $c_{y^*}(i^y) = 0$, we have $e_{c_y}(i^y, 1) \neq e_{c_{y^*}}(i^y, 1)$ and $e_{c_y}(i^y, 0) \neq e_{c_{y^*}}(i^y, 0)$. Therefore, $e_{c_y}$ has different outputs compared with $e_{c_{y^*}}$ at exactly four points, i.e. $\{(i^y, 1), (i^y, 0), (i^{y^*}, 1), (i^{y^*}, 0)\}$. This indicates $\forall e_{c_y} \in E_C$ with $y \neq y^*$,

$$
\begin{aligned}
&\text{er}_{(i,o)\sim I\times O}(e_{c_y}, e_{c_{y^*}}) \\
&\quad = \Pr_{(i,o)\sim I\times O}[e_{c_y}(i,o) \neq e_{c_{y^*}}(i,o)] \\
&\quad = \Pr_{(i,o)\sim I\times O}\left[(i,o) \in \left\{(i^y,0),(i^y,1),\left(i^{y^*},0\right),\left(i^{y^*},1\right)\right\}\right] \\
&\quad = \Pr_{i\sim I}\left[i = i^y \vee i = i^{y^*}\right] = \Pr_{i\sim I}\left[i = i^y\right] + \Pr_{i\sim I}\left[i = i^{y^*}\right].
\end{aligned} \tag{3.3}
$$

Note when $y = y^*$, $\text{er}_{(i,o)\sim I\times O}(e_{c_y}, e_{c_{y^*}}) = 0$. Therefore,

$$
\begin{aligned}
E_\epsilon &= \{e_{c_y} \in E_C : \text{er}_{(i,o)\sim I\times O}(e_{c_y}, e_{c_{y^*}}) \leq \epsilon\} \\
&= \left\{e_{c_y} \in E_C : \Pr_{i\sim I}[i = i^y] + \Pr_{i\sim I}\left[i = i^{y^*}\right] \leq \epsilon\right\} \cup \{e_{c_{y^*}}\} \\
&= \left\{e_{c_y} \in E_C : \Pr_{i\sim I}[i = i^y] \leq \epsilon - \Pr_{i\sim I}\left[i = i^{y^*}\right]\right\} \cup \{e_{c_{y^*}}\}.
\end{aligned} \tag{3.4}
$$

Because $\forall e_{c_y} \in E_\epsilon$, where $y \neq y^*$, is different from $e_{c_{y^*}}$ at exactly four points, we have

$$
\begin{aligned}
\text{DIS}(E_\epsilon) = &\left\{(i,o) \in I \times O : \Pr_{i\sim I}[i = i^y] \leq \epsilon - \Pr_{i\sim I}\left[i = i^{y^*}\right],\right. \\
&\left. o \in \{0,1\}\right\} \cup \left\{\left(i^{y^*},1\right),\left(i^{y^*},0\right)\right\}.
\end{aligned} \tag{3.5}
$$

For tree inputs, if the logic values follow independent Bernoulli distribution with probability of 0.5, $\forall i^y \in I$, we have

$$\Pr_{(i,o)\sim I\times O}[(i,o)\in\{(i^y,0),(i^y,1)\}] = \Pr_{i\sim I}[i=i^y] = \frac{1}{2^n},$$

and $\forall e_{c_y} \in E_C$ with $y \neq y^*$,

$$\mathrm{er}_{(i,o)\sim I\times O}(e_{c_y}, e_{c_{y*}}) = \frac{1}{2^{n-1}}.$$

Therefore, by setting $\epsilon = 1/2^{n-1}$, we have $E_\epsilon = E_C$ and $\mathrm{DIS}(E_\epsilon) = I \times O$. According to the definition of $\theta$,

$$\theta = \frac{\Pr_{(i,o)\sim I\times O}[(i,o)\in \mathrm{DIS}(E_\epsilon)]}{\epsilon} = 2^{n-1}.$$

Hence proved.

Base on Lemma 3.1, we now have the following theorem concerning the security of a camouflaged AND-tree structure.

**Theorem 3.1** *For an n-input AND-tree structure with all tree inputs camouflaged with XOR-type camouflaging BUF cells, if the logic values for tree inputs follow identical independent Bernoulli distribution with probability of* 0.5*, then,*

$$m(e_{c_{y*}}, E_C) = \mathcal{O}(2^n).$$

***Proof 3.2*** Based on Lemma 3.1, we have $\theta = 2^{n-1}$ for an $n$-input AND-tree. Meanwhile, because $|E_C| = 2^n$, we have $d \geq \log_2 |E_C|/n = 1$. Therefore, $m(e_{c_{y*}}, E_C) = \mathcal{O}(2^n)$. Hence proved.

From Theorem 3.1, under the assumption that the logic values for tree inputs follow identical independent Bernoulli distribution with probability of 0.5, we can formally prove the security of an $n$-input AND-tree by showing that the de-camouflaging complexity of a SAT-based attack scales exponentially with the increase of tree input size.

### AND-Tree Structure in General Circuits

According to the analysis above, a stand-alone AND-tree structure can lead to exponential increase of de-camouflaging complexity. However, this may not be true for an AND-tree structure in general circuits due to the following reasons:

- The input pins to the AND-tree structure may not be primary inputs to the circuit. As shown in Fig. 3.9a, since the fanin cone of different input pins can be overlapped, the requirement on independence may not be satisfied. Meanwhile, depending on the logic gates in the fanin cone, the signal probability for each tree input may also deviate from 0.5.

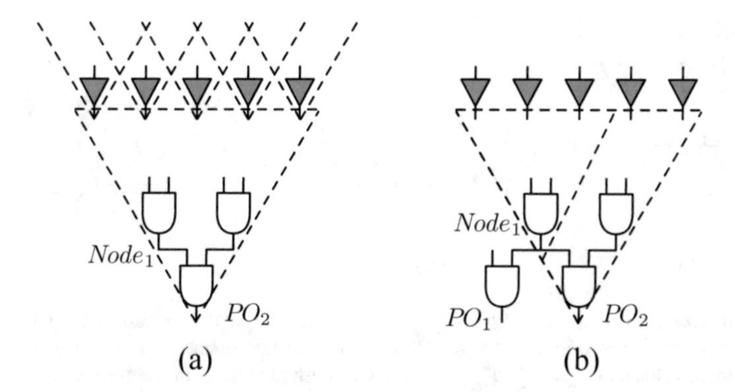


**Fig. 3.9** Two situations that can impact the security of AND-tree structure: (**a**) overlapped fanin cone for input pins leads to correlation; (**b**) extra path to primary outputs from internal node makes it possible to decompose the tree

- There are usually more than one primary outputs in the circuit and more than one paths from some internal nodes of an AND-tree to the primary outputs. Consider the AND-tree structure in Fig. 3.9b. The internal node $Node_1$ can bypass the root of the tree $PO_2$ and get observed at the primary output $PO_1$. This can also reduce the de-camouflaging complexity of the AND-tree structure.

Therefore, to determine the security of an AND-tree structure in general circuits, we need to characterize the two factors, which we define as input bias and tree decomposability.

## Input Bias Evaluation

Input bias is proposed to characterize the distance between actual joint distribution for logic values at tree input pins and the ideal independent Bernoulli distribution.

As shown in Fig. 3.9a, the logic value of input pins is determined by the primary inputs and logic gates in the fanin cones, which makes the original assumption on independent Bernoulli distribution for input pins invalid. We denote this as input bias since the actual input distribution deviates from the ideal distribution. Input bias mainly impact $E_\epsilon$ and $\mathrm{DIS}(E_\epsilon)$. According to Eq. (3.5), to decide $\mathrm{DIS}(E_\epsilon)$, we need to calculate the probability of each input vector, which, however, is intractable for large circuits. To capture the impact of input bias, we instead consider the following approximate approach.

According to Eq. (3.3), $\forall y \neq y^*$, we have

$$\mathrm{er}_{(i,o)\sim I\times O}(e_{c_y}, e_{c_{y*}}) = \Pr_{i\sim I}[i = i^y] + \Pr_{i\sim I}\left[i = i^{y^*}\right] \geq \Pr_{i\sim I}\left[i = i^{y^*}\right].$$

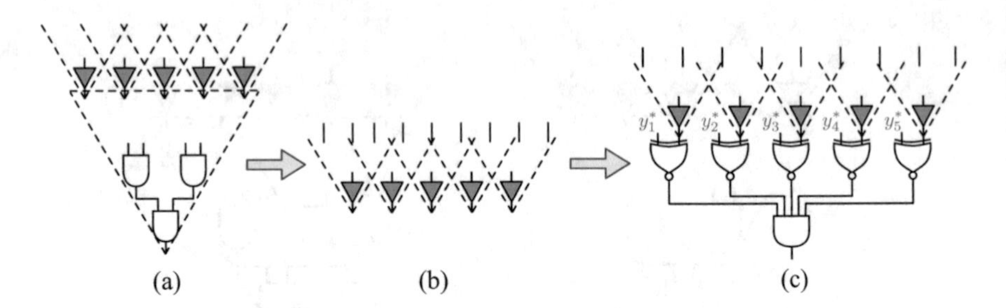


**Fig. 3.10** Exact calculation of $\Pr_{i\sim I}[f_{y^*}(i) = 1]$: (**a**) original tree structure; (**b**) fanin cone extracted for all the tree input pins; (**c**) form the circuit that connects all tree input pins to the desired logic values, i.e. $\{y_1^*, y_2^*, y_3^*, y_4^*, y_5^*\}$. By forcing the output of the formed circuit to 1, we can solve the logic values for the circuit primary inputs iteratively as in Algorithm 1

To get a non-empty $E_\epsilon$, we must choose $\epsilon \geq \Pr_{i\sim I}[i = i^{y^*}]$. Because we always have $\Pr_{(i,o)\sim I\times O}[(i,o) \in \text{DIS}(E_\epsilon)] \leq 1$, then

$$\theta = \frac{\Pr_{(i,o)\sim I\times O}[(i,o) \in \text{DIS}(E_\epsilon)]}{\epsilon} \leq \frac{1}{\Pr_{i\sim I}[i = i^{y^*}]}.$$

Therefore, to evaluate the impact of input bias, we can first calculate $\Pr_{i\sim I}[i = i^{y^*}]$ to get the upper bound of $\theta$. If the upper bound is smaller than the pre-defined requirement, then, we conclude the AND-tree in the circuit is not enough to guarantee the security.

To evaluate $\Pr_{i\sim I}[i = i^{y^*}]$, we consider the procedure as shown in Fig. 3.10. We first extract the fanin cone for all the tree input pins as in Fig. 3.10b. Then, as in Fig. 3.10c, we form the circuit that connects each tree input pin to its desired logic value. The output of the formed circuit equals to 1 if and only if the logic values for all the tree inputs equal to $y^*$. Therefore, if we force the output equal to 1 and solve the value for the circuit primary inputs, we can get $\Pr_{i\sim I}[i = i^{y^*}]$. We formulate the problem into a SAT problem and show the pseudo code in Algorithm 1. FORMSATPROB grabs the fanin cone for the tree input pins and forms the SAT equation as in Fig. 3.10c (line 1). *Cnt* is used to count the total number of input vectors that satisfy the SAT equation, which in turn can be used to calculate $\Pr_{i\sim I}[i = i^{y^*}]$. We initialize *Cnt* to 0 and iteratively solve the SAT problem to search for the input vectors until the SAT problem is not satisfiable. As we have discussed, to guarantee security, $\Pr_{i\sim I}[i = i^{y^*}]$ cannot be larger than a threshold. This enables us to set a threshold for the maximum value of *Cnt*, i.e. $T_h$ in line 6. The efficiency highly depends on the $\Pr_{i\sim I}[i = i^{y^*}]$ as it determines the iterations of the algorithm. Because to guarantee the security level, $\Pr_{i\sim I}[i = i^{y^*}]$ is usually small, the overall efficiency of the algorithm is quite acceptable.

Given $\Pr_{i\sim I}[i = i^{y^*}]$, we can determine the upper bound of the de-camouflaging complexity for the tree structure. When the upper bound is large enough, to further evaluate the tree structure, we calculate the distance between the actual distribution

**Algorithm 1** Algorithm of calculating $\Pr_{i \sim I}[i = i^{y^*}]$

1: $F \leftarrow$ FORMSATPROB$(G, i, y, y^*)$;
2: $Cnt \leftarrow 0$;
3: **while** $F$ is satisfiable **do**
4: $\quad i_t \leftarrow$ SATSOLVE$(F)$;
5: $\quad Cnt \leftarrow Cnt + 1$;
6: $\quad$ **if** $\frac{Cnt}{2^{|i|}} \geq T_h$ **then**
7: $\quad\quad$ **Return** $\frac{Cnt}{2^{|i|}}$;
8: $\quad$ **end if**
9: $\quad F \leftarrow F \wedge (i \neq i_t)$;
10: **end while**
11: **Return** $\frac{Cnt}{2^{|i|}}$;

for logic values of tree input pins compared with the ideal distribution. Our intuition is that while the ideal distribution provides the best security, the closer the actual distribution is compared to the ideal distribution, the better security the tree structure can provide. To evaluate the distance between different distributions, we adopt the normalized Kullback–Leibler (KL) divergence [16].

Normalized KL divergence for two discrete probability distributions, $P$ and $Q$, is calculated as

$$KL(P|Q) = \frac{1}{n}\sum_{i} P(i)\log\frac{P(i)}{Q(i)}. \tag{3.6}$$

In our case, since $Q$ is uniform, $KL(P|Q) = (n - H_p)/n$, where $H_p$ is the total entropy of distribution $P$. Note that the larger the KL divergence is, the closer $KL(P|Q)$ approaches to 1 and the worse the security of the AND-tree is.

In summary, the strategy to evaluate the input bias of existing tree structures in the original circuits becomes

- First, evaluate $\Pr_{i \sim I}(i = i^{y^*})$ following Algorithm 1 to determine the upper bound of $\theta$.
- Second, when the upper bound is large enough, do random sampling for circuit primary inputs and then, derive the logic value for the tree input pins, based on which, $KL(P|Q)$ can be calculated following Eq. (3.6). If $KL(P|Q)$ is smaller than a pre-defined threshold, then, we consider the provided security of the tree structure to be large enough.

#### Tree Decomposability Characterization

To characterize the impact of multiple paths to primary outputs, we propose the concept on tree decomposability.

**Algorithm 2** Determine whether an AND-tree is decomposable

1: // $G$ is the original circuit and $G_t$ is the AND-tree
2: // $r_t$ is the root of the tree
3: $U \leftarrow$ TOPOLOGICALSORT($r$, $G_t$);
4: **for** $u \in U$ **do**
5:     **if** $u$.fanout $> 1$ **then**
6:         $\{p_1, \ldots, p_m\} \leftarrow$ DFS($u$, $G$);
7:         **if** $\exists i$, s.t. $r \notin p_i$ **then**;
8:             **return** True;
9:         **end if**
10:     **end if**
11: **end for**
12: **return** False;

**Definition 3.1 (Decomposable Tree)** An AND-tree structure is decomposable if (1) there exists a path from the internal node of the tree to the primary output that can bypass the root of the tree and (2) change of the logic value of the internal node can be observed at the primary output through the path.

Both of the conditions are important. For example, the AND-tree structure in Fig. 3.9b is decomposable because the internal node $Node_1$ can bypass the root of the tree $PO_2$ and get observed at the output $PO_1$. Tree decomposability is undesired because it enables the attacker to first de-camouflage the sub-tree structure rooted at $Node_1$, and then de-camouflage the remaining part of the tree, which is also an AND-tree structure, but with fewer input pins. The number of input vectors needed to de-camouflage the decomposable AND-tree is thus limited to the sum of the input vectors needed to de-camouflage the two subtrees. Due to tree decomposability, the size of the two subtrees is much smaller than the original AND-tree, which indicates much smaller de-camouflaging complexity.

To determine whether an AND-tree is decomposable, we propose the algorithm shown in Algorithm 2. We traverse the tree structure in a reverse topological order starting from the root. For each internal node $u$ of the tree, if it has more than 1 successors, then, we do a depth-first search starting from $u$ and keep record of all the paths from $u$ to the primary outputs. If the tree root exists in each path, then, the tree is non-decomposable.

### *3.3.5 Provably Secure IC Camouflaging*

In this section, we will leverage the proposed camouflaging cell generation method and the AND-tree structure to provide provably secure camouflaging strategy. The overall flow of the proposed IC camouflaging framework is illustrated in Fig. 3.11. The first step is the camouflaging cell library generation with the proposed techniques described in Sect. 3.3.3. Then, accurate characterization is performed to determine the timing, power, and area overhead for each cell in the camouflaging

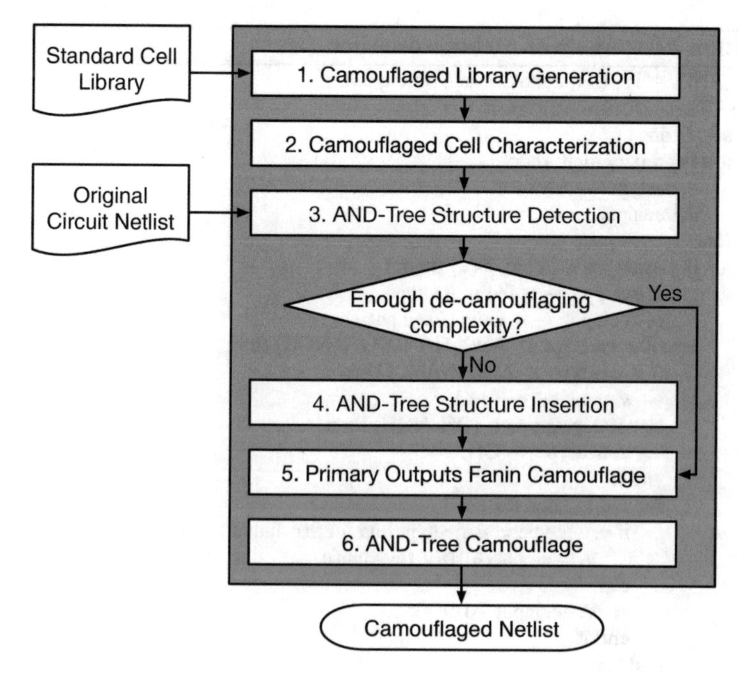


**Fig. 3.11** The proposed IC camouflaging flow

cell library. In the third step, existing AND-tree structure is detected for the original netlist. If the pre-defined de-camouflaging complexity is not satisfied, new AND-tree structure needs to be inserted as in the fourth step. Otherwise, we can simply camouflage the AND-tree structure to enforce the resilience to structural attacks. In the fifth step, we leverage the inserted AND-tree structure to protect all the primary outputs to ensure that at least one large AND-tree exists in the fanin cone of each primary output. We further camouflage the AND-tree structure to enhance the resilience against tree removal attack in the sixth step. After the sixth step, a camouflaged netlist will be generated.

### AND-Tree Detection in Original Netlist

AND-tree represents a set of circuit structures. We denote all the circuit structures that generate 1 as output for only one input vector as AND-tree and those that generate 0 as output for only one input vector as OR-tree. The pseudo code of the algorithm we propose to detect the tree structure is shown in Algorithm 3. We start from the primary inputs of the circuit and sort all the circuit nodes in a topological order (line 2). For each node, we keep record of the tree rooted at this node by recording the input pins of the tree. For primary inputs, the type of tree rooted at the node can be treated as either AND-type or OR-type (lines 4–6). For the internal

**Algorithm 3** Algorithm of And-tree detection

```
 1: Let {AND, ANY, OR} denote a set of tree types.
 2: U ← TOPOLOGICALSORT(G);
 3: for u ∈ U do
 4:     if u is primary input then
 5:         u.treetype ← ANY;
 6:         u.treeinput ← u;
 7:     else
 8:         if u.gatetype ∈ {BUF, INV} then
 9:             u.treetype ← u.fanin.treetype;
10:             u.treeinput ← u.fanin.treeinput;
11:         else if u.gatetype ∈ {AND, NAND, OR, NOR} then
12:             if u.gatetype ∈ {AND, NAND} then
13:                 u.treetype ← AND;
14:             else if u.gatetype ∈ {OR, NOR} then
15:                 u.treetype ← OR;
16:             end if
17:             for v ∈ u.fanin do
18:                 if v.treetype = u.treetype and SIZE(v.fanout) = 1 then
19:                     u.treeinput.ADD(v.treeinput);
20:                 else
21:                     u.treeinput.ADD(v);
22:                 end if
23:             end for
24:         else
25:             u.treetype ← ANY;
26:             u.treeinput ← u;
27:         end if
28:         if u.gatetype ∈ {INV, NOR, NAND} then
29:             u.treetype ← INVERT(u.treetype);
30:         end if
31:     end if
32: end for
33: return U.
```

nodes, to determine the input pins of the tree structure, we consider the gate type of the node and its predecessors in the circuit graph. Depending on the type of the gate, there are following possibilities (lines 7–31):

- If the gate is INV or BUF, the node will have the same tree as its input (lines 8–10). For INV, function INVERT() is called to change the tree type from AND-type to OR-type or vice versa.
- If the gate is AND or OR, the tree type rooted at the node can first be determined (lines 12–16). Then, to determine the input pins, there are two possible situations depending on the predecessors' tree types and the tree type of the node. When the tree types are the same, larger tree structure can be formed (lines 18–19). In this case, function ADD() is called to add the predecessor's tree structure to the node. When the tree types are different, only the predecessor itself can be added to the tree (lines 20–22).

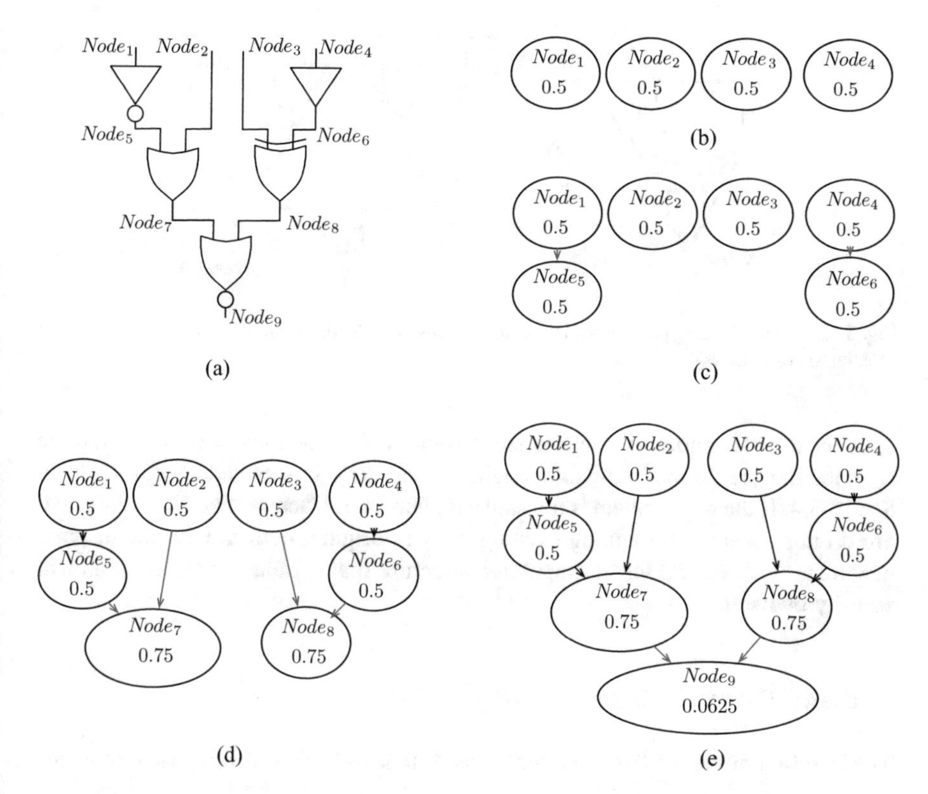


**Fig. 3.12** Detect tree structure in topological order: (**a**) circuit netlist; (**b**)–(**e**) start from primary inputs to calculate tree structure in topological order

- If the gate is NAND or NOR, we can treat it as an AND or OR connected with INV and follow the procedure above.
- For other type of gates, including XOR, XNOR, MUX, and so on, no tree structure can be formed and the node itself is added to the tree (lines 24–27).

We now use an example to illustrate the Algorithm 3.

*Example 3.4* Consider the circuit shown in Fig. 3.12a. As in Fig. 3.12b, for primary input $Node_1$, the tree type is ANY and the input of the tree is $\{Node_1\}$. For internal node $Node_5$, since it is connected with $Node_1$ through a BUF, the tree type for $Node_5$ is also ANY and the input is also $\{Node_1\}$. For $Node_8$, since it is connected with an XOR gate, the tree type becomes ANY and the input to the tree is the node itself, i.e. $\{Node_8\}$. Consider $Node_7$, since it is connected with an OR gate, the tree type has to be OR. For the two inputs, i.e. $Node_5$ and $Node_2$, since the tree types for both nodes are ANY, they can be combined to form large tree structure. Therefore, the input pins for the tree rooted at $Node_7$ become $\{Node_1, Node_2\}$. Similarly, we can determine the tree type and tree inputs for $Node_9$. Because $Node_9$ is connected with an NOR gate, the tree type becomes AND.

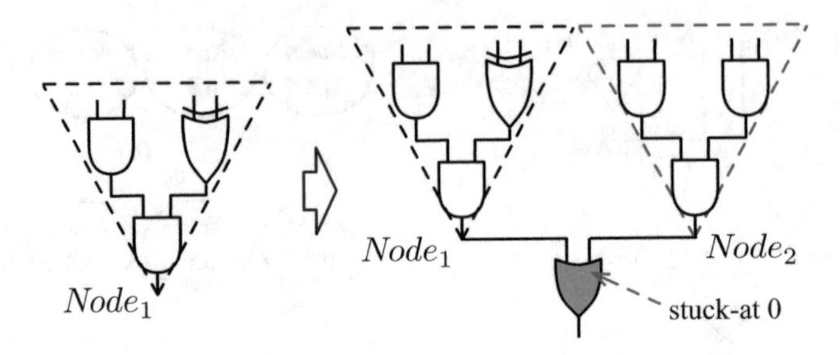


**Fig. 3.13** Insert AND-type tree structure to the circuit (*Node*$_2$ is stuck-at-0 to guarantee the functional correctness)

After the calculation of the tree structure rooted at each node, we can examine whether the pre-defined de-camouflaging complexity is satisfied as described in Sect. 3.3.4. If the requirement is not satisfied, new tree structures need to be inserted. We do not consider combining existing trees in original netlists with the inserted new trees. Instead, we insert a new tree structure that is able to provide sufficient security by itself.

### Stochastic Greedy AND-Tree Insertion

The insertion of the AND-tree structure needs to satisfy the following requirements:

- The functionality of the original circuit is not changed.
- The overhead induced by the insertion should be minimized.
- A false interpretation of the AND-tree functionality leads to erroneous outputs.

To satisfy the first requirement, we leverage the STF-type camouflaging cells as described in Sect. 3.3.3. Consider an example circuit as shown in Fig. 3.13. To insert an AND-tree structure at *Node*$_1$, we first insert an OR gate to *Node*$_1$ with the other input *Node*$_2$ as dummy pin. Then, an AND-tree structure is created with *Node*$_2$ being the root. The input pins of the AND-tree structure are connected to the primary inputs and camouflaged with XOR-type cells.

To detect the stuck-at-0 fault at *Node*$_2$, we again follow the same analysis as in Sect. 3.3.4. The logic value of *Node*$_2$, which is 0 in reality, can be expressed as

$$c_y(i) = g_{n+1}(g_1(i_1) \wedge g_2(i_2) \wedge \ldots \wedge g_n(i_n)).$$

Note that $g_{n+1}(i) = 0$ indicates there is a stuck-at-0 fault at $Node_2$, and $g_{n+1}(i) = i$ otherwise. Among all the possible configurations, there are $2^n$ correct configurations with $g_{n+1}$ interpreted as stuck-at-0 and $2^n$ incorrect configurations with $g_{n+1}(i) = i$. For any false configuration $y$, $c_y$ outputs 1 for exactly one input vector, denoted as $i^y$, and, thus, is different from $c_{y^*}$ for exactly one input vector. For the corresponding indicator function, $e_{c_y}$ is different from $e_{c_{y^*}}$ at exactly two points, i.e. $\{(i^y, 1), (i^y, 0)\}$. Therefore, $\forall y$ with $c_y \neq c_{y^*}$, we have

$$\begin{aligned}
&\text{er}_{(i,o)\sim I\times O}(e_{c_y}, e_{c_{y*}}) \\
&= \Pr_{(i,o)\sim I\times O}[e_{c_y}(i,o) \neq e_{c_{y*}}(i,o)] \\
&= \Pr_{(i,o)\sim I\times O}[(i,o) \in \{(i^y, 0), (i^y, 1)\}] \\
&= \Pr_{i\sim I}[i = i^y].
\end{aligned} \tag{3.7}$$

Since we connect the input pins of the inserted tree structure with circuit primary inputs, we can assume no input bias for tree inputs, which indicates

$$\Pr_{i\sim I}[i = i^y] = \frac{1}{2^n}.$$

Similar to proof in Sect. 3.3.4, if we set $\epsilon = \frac{1}{2^n}$, then, we have $E_\epsilon = E_C$ and $\text{DIS}(E_\epsilon) = I \times O$. In this case, $\theta = 2^n$. Therefore, $m(e_{c_{y*}}, E_C) = \mathcal{O}(2^n)$.

Therefore, the required number of input vectors to de-camouflage the circuit increases exponentially to the size of the inserted AND-tree structure. The insertion of OR-tree follows the same procedure except that we need to use an AND gate with stuck-at-1 fault at the dummy input, which is the root of the OR-tree structure.

To determine the location for the insertion of the tree structure, we propose a stochastic greedy tree insertion algorithm as shown in Algorithm 4, which tries to minimize the performance overhead and guarantee the functionalities for all primary outputs are protected. We first add all the primary outputs that we hope to protect in a set $U_{PO}$. Then, to decide the candidate circuit node for tree insertion, we traverse the circuit graph in a topological order and calculate an insertion score (*IS*) for each internal node. *IS* is defined to consider the node's switching probability *SA*, observe probability $P_{ob}$ and the number of primary outputs $N_O$ that have not been camouflaged in its fanout cone, which is calculated as

**Algorithm 4** Algorithm of stochastic greedy AND-tree insertion

1: $U_{PO} \leftarrow POs$;
2: $U \leftarrow$ TOPOLOGICALSORT($G$);
3: REMOVECRITICALNODE($U$);
4: **while** $U_{PO} \neq \emptyset$ **do**
5:     **for** $u \in U$ **do**
6:         $u.\text{score} \leftarrow$ COMPUTEIS($G$);
7:     **end for**
8:     $U_{IS} \leftarrow$ FINDTOPK($U$);
9:     $u_c \leftarrow$ RANDOMSELECTCAND($U_{IS}$);
10:     $G \leftarrow$ ANDTREEINSERT($u_c$, $G$);
11:     $U_{PO} \leftarrow$ REMOVECOVEREDPO($U_{PO}$, $u_c$);
12: **end while**
13: **return** $U$;

$$IS = \frac{\alpha \times SA - \beta \times P_{ob}}{N_O}. \tag{3.8}$$

By defining *IS* following Eq. (3.8), we look for the circuit node with lowest average cost to camouflage one primary output. Here, cost is defined considering introduced power overhead $SA$ and error observability $P_{ob}$. $\alpha$ and $\beta$ are coefficients defined to balance $SA$ and $P_{ob}$. By increasing $\alpha$, we can reduce the introduced power overhead, while by increasing $\beta$, we prefer circuit nodes which lead to better error probability at primary outputs. Note that before we calculate the score, all the circuit nodes along timing critical paths are removed first to guarantee negligible impact on performance. Then, we find $k$ nodes with smallest scores from $U$ and randomly select one node as the candidate for tree insertion. All the primary outputs in the fanout cone of the candidate node are removed from $U_{PO}$ and the procedure is continued until all the primary outputs are protected.

The insertion of AND-tree structure helps camouflage the functionality of original netlist. While the timing overhead can be small since the nodes along critical paths are not changed, the introduced power and area overhead cannot be avoided. However, the size of the inserted tree only depends on the required security level and is independent of the size of original netlist. Meanwhile, while the induced area and power overhead increases linearly as the tree size, the de-camouflaging complexity increases exponentially. Therefore, to ensure certain de-camouflaging complexity, the overhead is acceptable. For relatively large circuit, the overhead is even negligible.

More importantly, the de-camouflaging complexity, which is defined as the number of input–output patterns required for de-camouflaging, is independent of the way that a SAT-problem is formulated and the software package or computer configuration that the attack is carried on. Therefore, the proposed camouflaging framework is provably secure provided that the requirement on the size of non-decomposable tree and input bias is satisfied.

#### AND-Tree Camouflaging Against Removal Attack

By inserting AND-tree into original circuit netlists, the de-camouflaging complexity can be increased exponentially, which ensures good resilience against SAT-based attack. However, because large AND-tree is a unique structure that does not usually exist in general circuit netlist, it is possible for the attacker to identify and remove it. In [46], the authors propose to identify the AND-tree by calculating the signal probability skew (SPS) for each circuit node. SPS of a signal $s$ is defined as $\Pr[s = 1] - 0.5$. In Fig. 3.14, we use an example to illustrate the attack process. Starting from primary inputs, the attacker traverses the circuit netlist topologically. For a standard cell, the signal probability can be easily calculated. However, for a camouflaging cell, because its actual functionality in the circuit is unknown to the attackers, the same probability is assumed for each functionality. Therefore, the output signal probability is computed as the average of the output signal probability

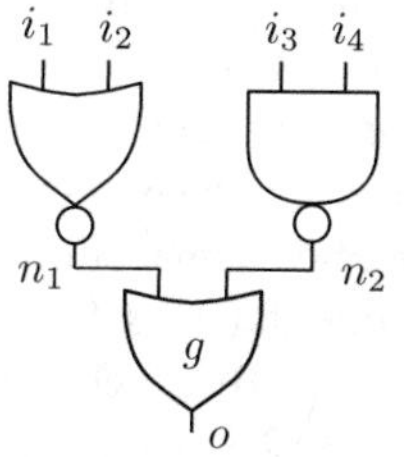

$$SPS(i_1) = SPS(i_2) = SPS(i_3) = SPS(i_4) = 0$$

$$SPS(n_1) = \frac{1}{2} \times \frac{1}{2} - \frac{1}{2} = -\frac{1}{4}, SPS\ (n_2) = 1 - \frac{1}{2} \times \frac{1}{2} - \frac{1}{2} = \frac{1}{4}$$

When g is OR gate:

$$SPS(o) = 1 - \frac{1}{4} \times \frac{3}{4} - \frac{1}{2} = \frac{5}{16}$$

When g is XOR-type OR gate:

$$SPS(o) = \frac{1}{2}(1 - \frac{1}{4} \times \frac{3}{4}) + \frac{1}{2}(\frac{1}{4} \times \frac{3}{4}) - \frac{1}{2} = 0$$

**Fig. 3.14** SPS-based functional attack: when $g$ is OR gate, $SPS(o) = \frac{5}{16}$ and when $g$ is XOR-type camouflaging cell, whether $g$ works as an OR gate or a NOR gate is equally possible for the attacker [46], and thus $SPS(o) = 0$

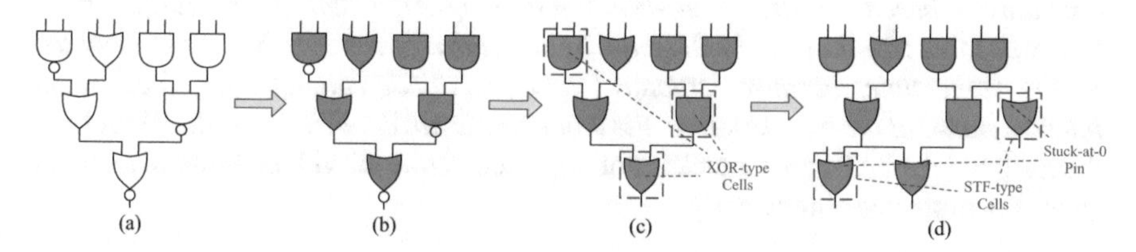


**Fig. 3.15** AND-tree protection: (**a**) original AND-tree structure; (**b**) and (**c**) functional camouflaging to reduce SPS for the root node; (**d**) structural camouflaging to add dummy connections from AND-tree internal nodes

of each possible cell functionality. For a signal with large uncertainty, its SPS tends to approach 0. However, for the root of an AND-tree, its SPS approaches to −0.5 exponentially with respect to the size of the AND-tree. This makes it possible for the attacker to identify the inserted tree structure by SPS.

Besides the SPS-based attack, because a non-decomposable AND-tree is an isolated structure that does not have many connections with the original circuit, the attackers can also leverage this structural footprint to detect the inserted AND-tree structure. To protect the inserted AND-tree from such removal attack, we propose to camouflage the structure both functionally and structurally.

We use the example in Fig. 3.15 to illustrate our AND-tree camouflaging strategy. Consider an 8-input AND-tree in Fig. 3.15a. We first replace the standard cells in the AND-tree with camouflaging cells that look the same and share the same functionality, as in Fig. 3.15b. Then, we replace the NAND, NOR, and INV cells with XOR-type camouflaging cells that look differently but share the same functionality, as in Fig. 3.15c. Because the attacker cannot determine whether the output of each cell in the AND-tree is negated or not, e.g. whether an AND cell works as an AND or a NAND cell in the circuit, the SPS for each node in the AND-tree is always kept as 0 according to [46]. Therefore, functional attacks by SPS are rendered useless.

To prevent removal attack based on structural information, we leverage the STF-type camouflaging cell to connect the internal nodes of the AND-tree to other gates as in Fig. 3.15d. For an $\tilde{n}$-input AND-tree, there are in total $\tilde{n} - 1$ gates in the tree following the structure in Fig. 3.15a. To ensure the size of the largest AND-tree detected by the attacker to be less than $\tilde{n}'$, we can always insert $O((\tilde{n} - 1)/(\tilde{n}' - 1))$ STF-type camouflaging cells to create dummy connections to the internal nodes as in Fig. 3.15d. Meanwhile, to prevent the attackers from identifying the inputs to the AND-tree, we can insert extra XOR-type BUF cells to other primary inputs. *Note that by inserting dummy connections with STF-type cells, the original circuit functionality is maintained. Meanwhile, the inserted AND-tree is still non-decomposable from the defense perspective since the logic value of AND-tree internal nodes cannot be sensitized through the dummy connections. However, from the attackers' point of view, because he cannot determine the connections are dummy based on structural attack following Algorithm 3, the largest non-decomposable tree that can be detected by structural attack is reduced significantly. At the same time, because structural and functional camouflaging focuses on internal nodes of the tree structure, the input bias is not impacted as well.* Therefore, the resilience to SAT-based attack is not impacted, while the vulnerability to removal attacks is mitigated significantly.

### Comparison Between State-of-the-Art Techniques

Until now, we have described our IC camouflaging strategy that detects, inserts, and camouflages AND-tree structure to provide guaranteed security towards SAT-based attack. Similar idea to leverage AND-tree structures has also been explored by Anti-SAT [40] and CamoPerturb [45] (Fig. 3.16). In this section, we compare the three strategies in terms of provided security, overhead, and their impact on the original circuit netlist, i.e. whether re-synthesis is required.

Assume an AND-tree with $\tilde{n}$-bit inputs is to be inserted into the circuit. According to Anti-SAT strategy, in fact, two subtrees, denoted as $Sub_1$ and $Sub_2$, are inserted into the circuit. $Sub_1$ and $Sub_2$ are AND-trees of the same size with $\tilde{n}$

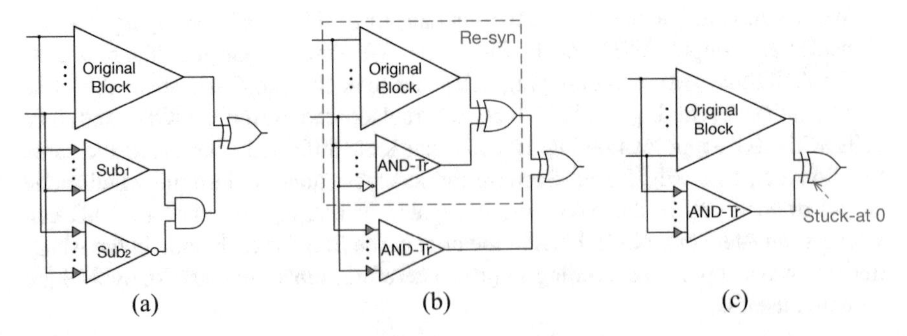


**Fig. 3.16** Comparison on AND-tree insertion strategy: (**a**) Anti-SAT [40]; (**b**) CamoPerturb [45]; (**c**) our insertion strategy

input pins and share the same input signals. An inverter is inserted at the output of $Sub_2$. XOR-type BUF cells are inserted at the input pins of the two AND-trees. For the circuit to function correctly, the XOR-type BUF cells of the same input signals in the $Sub_1$ and $Sub_2$ need to have the same functionality. To de-camouflage the circuit, the attacker always needs to query $2^{\tilde{n}}$ input vectors [40].

For CamoPerturb strategy, to insert an $\tilde{n}$-bit AND-tree, a specific input vector $i^*$ is first selected. Then, original circuit is re-synthesized by flipping the output value corresponding to $i^*$. An AND-tree is then inserted to correct the flipped output just for $i^*$. XOR-type BUF cells are inserted into the input pins of the AND-tree and their actual functionality in the circuit is determined by $i^*$. Based on [45], to de-camouflage the circuit, all input vectors are discriminating inputs and for each $i \neq i^*$, at most one false functionality can be pruned. However, it should be noted that $i^*$ can rule out all the false functionalities. Because $i^*$ is unknown to the attackers, on average, $2^{\tilde{n}-1}$ input vectors need to be measured. In the best case, the attacker has to measure $2^{\tilde{n}}$ input vectors to de-camouflage the AND-tree. However, in the worst case, only 1 input vector, i.e. $i^*$ is required for the attacker. It should be noted that for our AND-tree insertion strategy, as shown in Table 3.3, the de-camouflaging complexity is also $2^{\tilde{n}}$. This is because with our strategy, the circuit output is never impacted by the AND-tree output due to the stuck-at-0 input pin at the XOR gate. Therefore, there is no such input vector that is able to rule out all the incorrect circuit functionalities. Instead, for any input vector, exactly one false functionality can be ruled out.

The three strategies mainly introduce area and power overhead while the impact on timing can be negligible by avoiding any modification of circuit critical paths. Compared with our method, Anti-SAT suffers from almost double area and power overhead because one AND-tree and one NAND-tree of the same size are inserted. For CamoPerturb, besides the overhead introduced by the inserted AND-tree, extra overhead can be introduced in the process of re-synthesis. As we show in Sect. 3.3.6, large overhead can be introduced in the process of re-synthesis depending on $i^*$. We summarize the comparison in Table 3.3.

From the comparison we can see that our strategy for AND-tree insertion provides the best security guarantee. Meanwhile, because no re-synthesis is required for our methods, it introduces less modification to the original design and thus is easier for final design closure. We provide more detailed comparison on the introduced area and power overhead in experimental results.

**Table 3.3** Comparison with different tree insertion strategies

| Strategy | De-cam complexity | | | Overhead | Need Resyn |
|---|---|---|---|---|---|
| | Worst | Avg | Best | | |
| Ours | $2^{\tilde{n}}$ | $2^{\tilde{n}}$ | $2^{\tilde{n}}$ | $\tilde{n}$-bit AND-tree + $\tilde{n}$ cam. cells | No |
| CamoPerturb [45] | 1 | $2^{\tilde{n}-1}$ | $2^{\tilde{n}}$ | $\tilde{n}$-bit AND-tree + $\tilde{n}$ cam. cells + Resyn | Yes |
| Anti-SAT [40] | $2^{\tilde{n}}$ | $2^{\tilde{n}}$ | $2^{\tilde{n}}$ | $\tilde{n}$-bit AND-tree + $\tilde{n}$-bit NAND-tree + $2\tilde{n}$ cam. cells | No |

### 3.3.6 *Experimental Results*

In this section, we report on our experiments to demonstrate the effectiveness of the proposed IC camouflaging strategy. The camouflaging algorithm is implemented in C++. The SAT-based de-camouflaging algorithm is adopted from [34] and the SPS-based removal attack is implemented following [46]. We run all the experiments on an eight-core 3.40 GHz Linux server with 32 GB RAM. The benchmarks are chosen from ISCAS and MCNC benchmarks [4, 42]. For the de-camouflaging algorithm, we set the runtime limit to $1.5 \times 10^5$ s.

#### Verification of Camouflaging Cell Generation Strategy

We first demonstrate the security achieved by using camouflaging cell generation strategy. As described in Sect. 3.3.3, we first replace all the standard cells with camouflaging cells and then randomly change 10 cells with camouflaging cells that appear to be different but work with the same functionality. We show the introduced overhead, de-camouflaging complexity, and the time required for the SAT-based algorithm to resolve the original circuit functionality in Table 3.4. N/A indicates that the camouflaged netlist cannot be resolved within $1.5 \times 10^5$ s. As we can see, the area overhead, which is calculated as the sum of the area of each cell, is on average 0.68% and the power overhead is on average 0.55%, both of which are very small even for small benchmark circuits. Meanwhile, for large circuits, simply with the camouflaging cell generation strategy, the de-camouflaging algorithm cannot be finished within the pre-defined time. However, as we have pointed out in Sect. 3.3.3,

**Table 3.4** Verify the proposed camouflaging cell generation strategy by de-camouflaging complexity, time, attack on individual POs and introduced overhead

| Bench | | # input | # output | # gate | Time | # iter | Partial | Area (%) | Power (%) |
|---|---|---|---|---|---|---|---|---|---|
| ISCAS | c432 | 36 | 7 | 203 | 1.758 | 80 | 7/7 | 2.5 | 1.9 |
| | c880 | 60 | 23 | 466 | $1.2 \times 10^4$ | 148 | 23/23 | 1.1 | 0.85 |
| | c1355 | 41 | 32 | 619 | N/A | N/A | 29/32 | 0.86 | 0.73 |
| | c1908 | 33 | 25 | 938 | N/A | N/A | 0/25 | 0.58 | 0.71 |
| | c2670 | 233 | 64 | 1490 | N/A | N/A | 60/64 | 0.37 | 0.41 |
| | c3540 | 50 | 22 | 1741 | N/A | N/A | 9/22 | 0.27 | 0.13 |
| | c5315 | 178 | 123 | 2608 | N/A | N/A | 116/123 | 0.17 | 0.11 |
| MCNC | i4 | 192 | 6 | 536 | $1.9 \times 10^3$ | 743 | 6/6 | 1.2 | 0.94 |
| | apex2 | 39 | 3 | 652 | N/A | N/A | 1/3 | 0.77 | 0.62 |
| | ex5 | 8 | 63 | 1126 | $6.9 \times 10^2$ | 139 | 63/63 | 0.43 | 0.32 |
| | i9 | 88 | 63 | 1186 | $2.1 \times 10^4$ | 81 | 63/63 | 0.45 | 0.22 |
| | i7 | 199 | 67 | 1581 | $1.5 \times 10^2$ | 225 | 67/67 | 0.37 | 0.40 |
| | k2 | 46 | 45 | 1906 | N/A | N/A | 24/45 | 0.21 | 0.17 |
| | dalu | 75 | 16 | 2373 | N/A | N/A | 3/16 | 0.21 | 0.18 |

the SAT-based algorithm can still de-camouflage some small benchmarks with less than 1600 gates. Also, for the circuits that cannot be fully de-camouflaged, we can still run de-camouflaging attacks for each primary output separately and partially de-camouflage the design as shown in the partial column in Table 3.4. The experimental results demonstrate the effectiveness of the camouflaging cell generation strategy for large circuit, and also show the necessity to have other SAT-resilient protection strategies.

### Evaluation of AND-Tree Based Camouflaging Strategy

To evaluate the security of the AND-tree based camouflaging strategy, we start from stand-alone tree structures. We show the increase of the de-camouflaging complexity and time with respect to the tree size in Fig. 3.17a. As we can see, both the de-camouflaging time and complexity increase exponentially as we expect. To examine the impact of tree decomposability, we fix the size of an AND-tree,

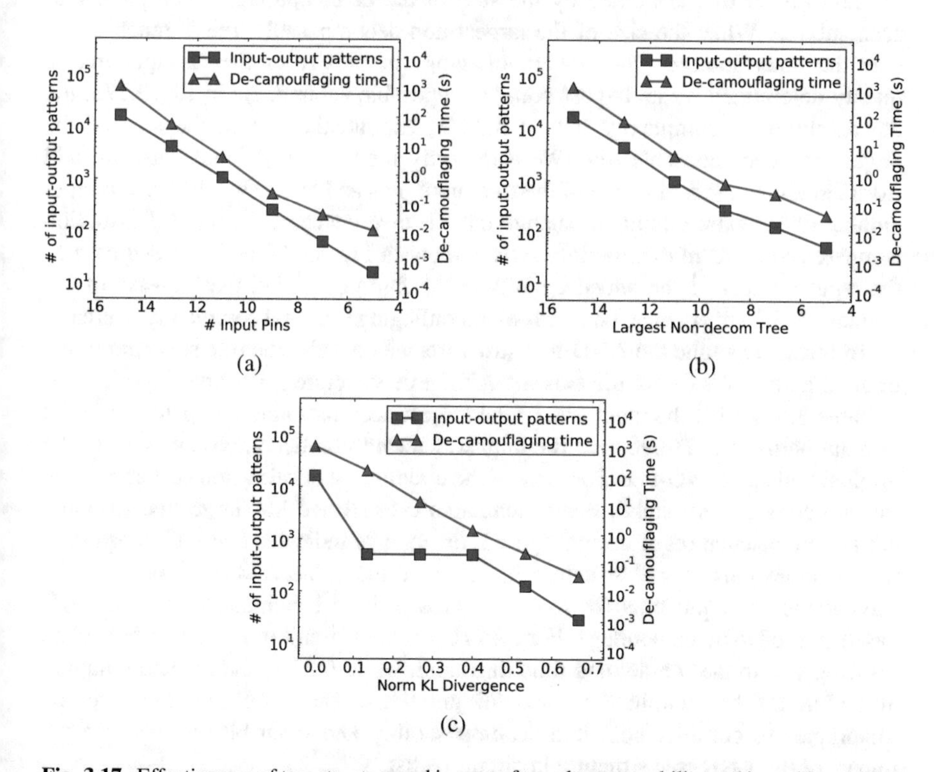


**Fig. 3.17** Effectiveness of tree structure and impact of tree decomposability and input bias: (**a**) de-camouflaging complexity and time for ideal AND-tree structure; (**b**) change of de-camouflaging complexity and time with the size of the largest non-decomposable tree; (**c**) change of de-camouflaging complexity and time with the input bias

**Table 3.5** Existing tree structure in benchmark circuits

| Bench | | D-tree | ND-tree | Norm KL div. |
|---|---|---|---|---|
| ISCAS | c1355 | 7 | 3 | 0.7 |
| | c1908 | 14 | 12 | 0.339 |
| | c2670 | 34 | 34 | 0.640 |
| | c3540 | 10 | 9 | 0.671 |
| | c5315 | 10 | 9 | 0.093 |
| MCNC | i4 | 7 | 7 | 0.732 |
| | ex5 | 6 | 6 | 0.589 |
| | i7 | 4 | 3 | 0.612 |
| | k2 | 175 | 39 | 0.968 |

i.e. 15 input pins, and change the size of the largest non-decomposable subtree in the 15-input AND-tree. The size of other non-decomposable subtrees is limited to be smaller than 3. We show the change of the de-camouflaging time and complexity in Fig. 3.17b. As we have discussed in Sect. 3.3.4, the de-camouflaging complexity of the 15-input tree is limited by the sum of the de-camouflaging complexity of each subtree. When the size of the largest non-decomposable tree is much larger than the other subtrees, the de-camouflaging complexity of the 15-input tree is mainly determined by its largest non-decomposable subtree. As in Fig. 3.17b, the de-camouflaging complexity indeed reduces exponentially with the size of the largest non-decomposable tree. We also verify the impact of input bias. We add extra circuits to the fanin cone of the tree input pins and gradually change the input number of the extra circuits to change the KL divergence of the input distribution compared to uniform distribution. As we show in Fig. 3.17c, with the decrease of the input number of the added circuits in the fanin cone, i.e. the increase of the normalized KL divergence, both the de-camouflaging time and complexity decrease.

To further examine the AND-tree structure, we consider the tree structure in the original netlist. We detect the existing AND-tree structure following Algorithm 3; In Table 3.5, we list the input size of the largest decomposable tree detected in the original netlist, i.e. D-tree, and the largest detected non-decomposable tree in the original netlist, i.e. ND-tree. For most of the circuits, the existing non-decomposable tree structure is very small. For benchmarks c2670 and k2, large tree structure exists. The calculation of normalized KL divergence indicates that high bias exists for the input pins of tree structure in k2 since the value is very close to 1. We camouflage the input pins for tree structures in both benchmarks and use SAT-based method to de-camouflage. For c2670, original circuit functionality cannot be resolved within the pre-defined time threshold, while for k2, the de-camouflaging algorithm finishes within 8.5 s and 70 iterations. The results demonstrate the importance to consider both tree decomposability and input bias to evaluate the impact of the AND-tree structure in circuit netlist.

Then, we insert tree structure into the benchmark circuits following Algorithm 4. We set $\alpha = \beta = 1$ for IS evaluation. We show the trade-off between the area overhead and the de-camouflaging time and complexity in Fig. 3.18 for benchmark

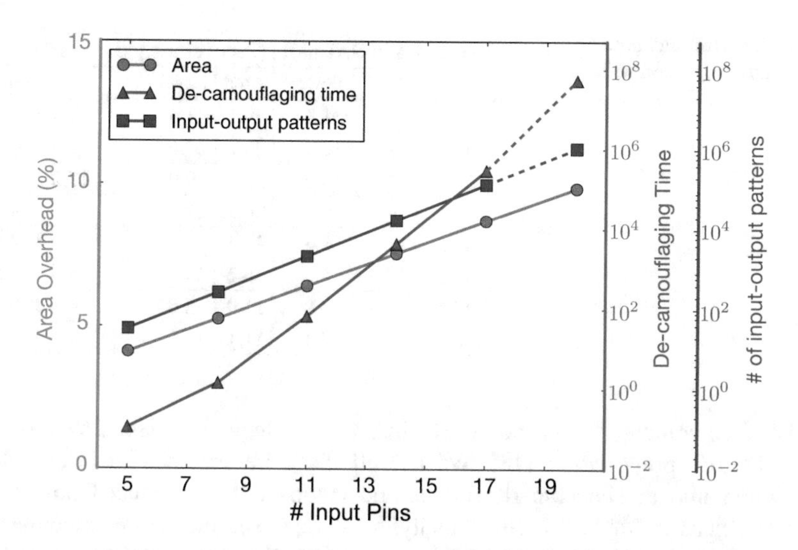


**Fig. 3.18** Trade-off between overhead and de-camouflaging complexity (dotted lines indicate extrapolation)

**Table 3.6** Introduced overhead of the tree-based camouflaging strategy when 64-input AND-tree is inserted

| Bench | Gate | Overhead | | | Security | |
|---|---|---|---|---|---|---|
| | | Area | Power | Timing | De-cam time | Partial |
| i4 | 536 | 32.5 | 19.4 | 0.5 | N/A | 0/6 |
| i9 | 1186 | 11.5 | 6.2 | 0.1 | N/A | 0/63 |
| c2670 | 1490 | 11.8 | 6.0 | 0.1 | N/A | 0/64 |
| i7 | 1581 | 9.7 | 4.7 | 0.2 | N/A | 0/67 |
| dalu | 2373 | 5.5 | 4.3 | 0.0 | N/A | 0/16 |
| c5315 | 2608 | 5.9 | 2.8 | 0.0 | N/A | 0/123 |
| c7552 | 3719 | 4.8 | 2.4 | 0.0 | N/A | 0/107 |
| des | 6729 | 1.9 | 1.2 | 0.0 | N/A | 0/245 |

c880. As we can see, the area overhead increases linearly with respect to the size of inserted tree while the de-camouflaging time and complexity increase exponentially. We then insert 64-input AND-tree structure to benchmark circuits. We leverage SAT-based attack to de-camouflage the camouflaged circuits. We also extract the subcircuits for each primary output and try to resolve the circuit separately. We show the results in Table 3.6. As we can see, SAT-based attack cannot de-camouflage any primary output of each benchmark circuit. We also report the introduced overhead, including power, area, and timing in Table 3.6. As shown in Table 3.6, the main overhead comes from area and power while the impact on timing is negligible. Meanwhile, for large circuit, e.g. des, the area and power overhead is less than 2%.

**Table 3.7** Area and power overhead comparison with Anti-SAT and CamoPerturb

| Bench | Anti-SAT [40] | | CamoPerturb [45] | | Ours | |
|---|---|---|---|---|---|---|
| | Area | Power | Area | Power | Area | Power |
| i4 | 62.2 | 38.9 | 49.4 | 41.6 | 32.5 | 19.4 |
| i9 | 23.4 | 12.3 | 29.1 | 29.6 | 11.5 | 6.2 |
| c2670 | 24.9 | 12.0 | 25.8 | 19.0 | 15.2 | 6.0 |
| i7 | 19.6 | 9.4 | 26.3 | 22.1 | 9.7 | 4.7 |
| dalu | 11.1 | 8.6 | 11.2 | 9.65 | 5.5 | 4.3 |
| c5315 | 12.6 | 5.6 | 16.2 | 13.3 | 7.9 | 2.8 |
| c7552 | 10.3 | 4.8 | 11.9 | 8.65 | 6.9 | 2.4 |
| des | 3.67 | 2.4 | 11.1 | 15.7 | 1.9 | 1.2 |

We then compare the proposed tree insertion strategy with Anti-SAT [40] and CamoPerturb proposed in [45]. We use all the three methods to insert 64-bit AND-tree into the benchmark circuits and compare the introduced power and area overhead in Table 3.7. Since analytical comparison on the de-camouflaging complexity is provided in Sect. 3.3.5, we do not run SAT-based attack for the three strategies. As in Table 3.7, our method achieves similar overhead compared with Anti-SAT. CamoPerturb suffers from larger power and area overhead compared with Anti-SAT and our strategy since large overhead is introduced in the re-synthesis process.

### Impact of Structural and Functional Camouflaging

We now verify the effectiveness of the structural and functional camouflaging for the AND-tree based camouflaging strategy and demonstrate the introduced overhead. We insert AND-tree with 64 input bins. We consider SPS-based methods proposed in [46] as functional attack and tree detection algorithm following Algorithm 2 as structural attack, which represents the state-of-the-art removal attack strategies. As shown in Table 3.8, after structural and functional camouflaging, the area and power overhead increases on average by 5.1 and 0.3%. However, for large benchmarks, i.e., des, the total area and power overhead after structural and functional camouflaging is less than 3%. After structural and functional camouflaging, SPS for the each internal node of AND-tree becomes 0.0, and the size of the largest non-decomposable AND-tree that can be detected following Algorithm 3 (i.e., "detected AND-tree" in Table 3.8) is 4. Therefore, structural and functional camouflaging can protect the inserted tree structure against the state-of-the-art removal attack strategies. Meanwhile, as we have discussed in Sect. 3.3.5, because the logic value of internal nodes of the AND-tree cannot be observed from the dummy connections, the overall resilience to SAT-attack is not reduced. As shown in Table 3.8, for the circuit netlists after structural and functional camouflaging, SAT-attack cannot be finished within pre-defined time threshold.

**Table 3.8** Verification of overhead and effectiveness of structural and functional camouflaging

| | ICCAD 2016 [20] | | | | | Ours | | | | |
|---|---|---|---|---|---|---|---|---|---|---|
| Bench | Area (%) | Power (%) | SPS | Detected ND-tree | SAT-attack | Area (%) | Power (%) | SPS | Detected ND-tree | SAT-attack |
| `i4` | 32.5 | 19.4 | 0.5 | 64 | N/A | 46.8 | 20.3 | 0.0 | 4 | N/A |
| `i9` | 11.5 | 6.2 | 0.5 | 64 | N/A | 16.4 | 6.5 | 0.0 | 4 | N/A |
| `c2670` | 15.2 | 6.0 | 0.5 | 64 | N/A | 19.1 | 6.3 | 0.0 | 4 | N/A |
| `i7` | 9.7 | 4.7 | 0.5 | 64 | N/A | 13.7 | 4.9 | 0.0 | 4 | N/A |
| `dalu` | 5.5 | 4.3 | 0.5 | 64 | N/A | 7.8 | 4.5 | 0.0 | 4 | N/A |
| `c5315` | 7.9 | 2.8 | 0.5 | 64 | N/A | 9.8 | 2.9 | 0.0 | 4 | N/A |
| `c7552` | 6.9 | 2.4 | 0.5 | 64 | N/A | 8.3 | 2.5 | 0.0 | 4 | N/A |
| `des` | 1.9 | 1.2 | 0.5 | 64 | N/A | 2.6 | 1.3 | 0.0 | 4 | N/A |

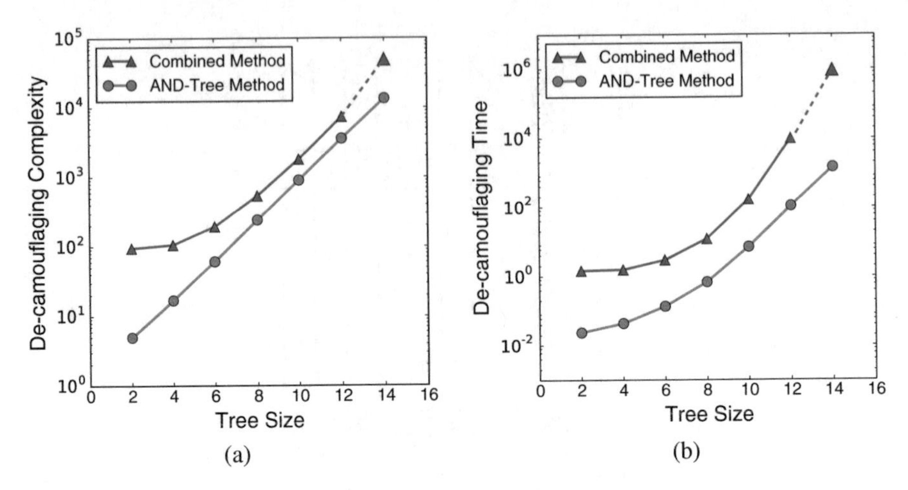


**Fig. 3.19** Comparison on the (**a**) de-camouflaging complexity and (**b**) de-camouflaging time between the combined techniques and the AND-Tree based technique on benchmark c880 (dotted line indicates extrapolation)

### Effectiveness of Combination of Two Camouflaging Strategies

Finally, we demonstrate the effectiveness of combining the two camouflaging strategies, i.e. camouflaging cell generation strategy and AND-tree based camouflaging strategy. To combine the two camouflaging strategies, we first insert an AND-tree structure into the original netlist following Algorithm 4. Then, we leverage the XOR-type and STF-type cells to further camouflage the circuit netlists following the strategy described in Sect. 3.3.3. We examine the effectiveness of the combined strategy by comparing the de-camouflaging complexity and time with the situation when only AND-tree based strategy is used. We run the experiments on benchmark `c880`. As shown in Fig. 3.19, by combining the two camouflaging strategies, both the de-camouflaging complexity and time are further increased, which indicates better security level.

## *3.3.7 Summary*

In this section, we have proposed a quantitative security criterion for de-camouflaging complexity measurements. The security criterion was formally analyzed based on the equivalence between the de-camouflaging strategy and the active learning scheme. Meanwhile, two camouflaging techniques were proposed: the low-overhead camouflaging cell library and the AND-tree structure, following the security criterion. A provably secure camouflaging framework was then developed to combine the two techniques, which achieves exponentially increasing security levels at the cost of linearly increasing overhead. Experimental results using the security criterion demonstrated that the camouflaged circuits with the proposed framework achieve high resilience against the SAT-based attack with only negligible performance overhead.

## 3.4 De-camouflaging Timing-Based Logic Obfuscation

To prevent AppSAT attacks [30], timing-based parametric camouflaging methods are proposed [41, 50]. Instead of directly hiding the circuit functionality, the parametric camouflaging methods obfuscate the timing behavior of the circuits. This is achieved by inserting buffers with tunable delay [41] or by introducing unconventional timing paths [50]. The obfuscated netlists consist of a combination of single-cycle paths and multi-cycle paths, which, if not resolved correctly, makes the reconstructed circuits malfunction. Therefore, the attackers are forced to determine the timing behavior for each path in the camouflaged netlists.

Timing-based approaches are promising and usually considered secure. Traditional testing-based methods that try to measure the path delay directly have been demonstrated impractical [50]. On the one hand, they suffer from poor scalability as the number of timing paths increases exponentially with respect to the circuit size. On the other hand, camouflaging techniques are proposed to hide selected timing paths and hinder the generation of the test vectors [50]. Timing-based approaches are also assumed to be secure against existing query-based attacks, e.g., SAT attacks. This is because existing query-based attacks focus on discriminating different Boolean functions and cannot be applied to determine the circuit timing schemes directly [41]. However, such an assumption remains unverified.

In Sect. 3.4, we formally study the resilience of the timing-based camouflaging strategies against query-based attacks. We propose the TimingSAT framework and demonstrate that after a proper transformation of the camouflaged netlist, existing SAT attacks are still effective to de-camouflage the timing-based strategies. To enable the transformation, we first come up with a novel key-controlled transformation unit (TU) design that is able to convert a timing path, either a single-cycle path or a multi-cycle path, into a single-cycle path without changing its functionality. The TUs are then inserted into the camouflaged netlist to convert all the paths into single-cycle paths with a guarantee that the correct circuit functionality can be achieved when proper keys are inserted.

After the transformation procedure, the de-camouflaging task is converted into determining the correct keys for the inserted TUs. Traditional SAT attacks cannot be directly applied due to the existence of uncontrollable flip flops in the inserted TUs. Therefore, we propose to unroll the transformed circuit for several time frames and apply the SAT attack to the unrolled netlist. The number of unrolling time frames is formally derived to guarantee the correctness of the resolved keys. While a direct implementation of TimingSAT can suffer from scalability issue for large benchmarks, we propose a simplification procedure to reduce the complexity of the unrolled netlist without compromising the correctness of the recovered keys. The performance of TimingSAT is validated with extensive experimental results. The contributions are summarized as follows:

- We propose a novel TU design and a transformation procedure to enable SAT attack on the timing-based camouflaging approaches.
- We formally prove the functional equivalence between the netlist recovered by TimingSAT and the original netlist.
- We propose a simplification procedure to significantly enhance the efficiency of TimingSAT without impacting the functionality of the recovered netlist.

The rest of Sect. 3.4 is organized as follows. Section 3.4.1 provides an overview of existing camouflaging techniques with an emphasis on the timing-based camouflaging approaches. Section 3.4.2 describes the proposed TU design and uses a motivating example to illustrate how it can help to determine the path timing schemes. Section 3.4.3 describes our TimingSAT algorithm, including the transformation and simplification techniques. Section 3.4.4 validates the performance of TimingSAT and investigates the runtime dependency of TimingSAT on different impacting factors. We conclude Sect. 3.4 in Sect. 3.4.5.

### *3.4.1 Preliminary: Timing-Based Camouflaging*

Timing-based parametric camouflaging approaches [41, 50] are described to prevent approximate SAT-based attacks. Besides regular camouflaging cells, [41] also inserts gates with tunable delay to the circuits. The delay configurations of these gates are carefully chosen to satisfy the pre-defined timing constraints and force the attackers to enumerate all the timing paths. Though the idea is promising, [41] suffers from one important drawback. Given the camouflaged netlist, the attackers can resolve the hold time constraints by assigning each gate a larger delay. Then, by operating the circuit with a lower frequency, the setup time constraints can be satisfied as well. In this way, the attackers can either operate the circuit correctly at a lower frequency or conduct re-synthesis to the resolved circuit to further optimize its performance.

The drawback is avoided in [50], which obfuscates the circuit timing by introducing multi-cycle paths. As shown in Fig. 3.20a, conventionally, all the paths in a combinational block operate within a single clock cycle. Zhang et al. [50] deliberately removes some flip flops, e.g., $F_2$ in Fig. 3.20a, to convert single-cycle paths into wave-pipelining paths. On a wave-pipelining path, e.g., the combinational path from $F_1$ to $F_3$ in Fig. 3.20b, there are more than one data waves propagating without a flip flop separating them. To retain the same functionalities as the original single-cycle circuit, the second wave cannot catch the first wave at any time during propagation as shown in Fig. 3.20c. Hence, the following timing constraints must be satisfied for all the wave-pipelining paths with two logic waves:

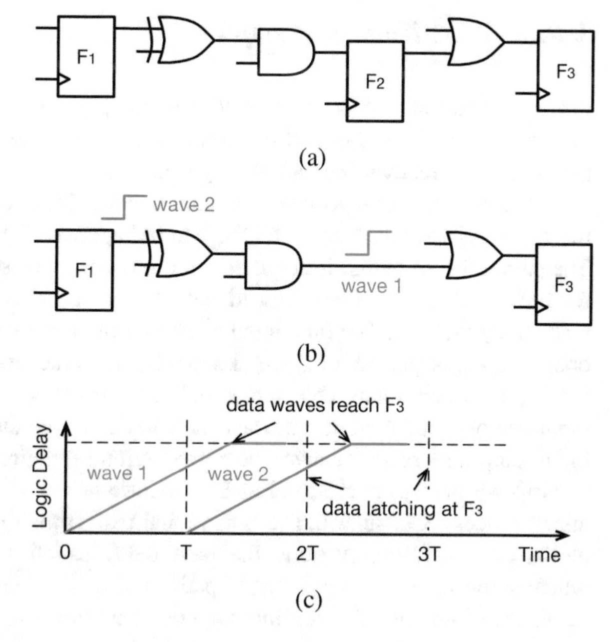


**Fig. 3.20** Conversion from single-cycle paths to wave-pipelining paths and the required timing constraints: (**a**) conventional paths with single-cycle timing scheme; (**b**) wave-pipelining paths with two logic waves; and (**c**) desired timing/spatial diagram for wave propagation on a wave-pipelining path

$$T_{clk} + t_h \leq d_p \leq 2T_{clk} - t_{su}, \tag{3.9}$$

where $d_p$ is the delay of the wave-pipelining paths and $T_{clk}$ represents the clock period time. $t_h$ and $t_{su}$ denote the hold and setup time for a flip flop.

By carefully removing the flip flops and sizing the gates, the functionality of the design remains unchanged with a mixture of single-cycle and wave-pipelining paths created. Due to the existence of wave-pipelining paths, the attackers cannot recover the correct circuit functionality by simply changing the clock period time. Therefore, they are forced to determine the timing scheme for each path to decide whether it is a single-cycle path or a wave-pipelining path.

To determine the path timing, physical reverse engineering alone is insufficient as it is usually difficult to measure the gate and interconnect delay accurately [50]. Assume the delay extraction techniques incur an inaccuracy factor $\tau$ $(0 < \tau < 1)$. For a path with delay $d_p$, if $d_p$ satisfies

$$(1 - \tau)d_p \leq T_{clk} \leq (1 + \tau)d_p, \tag{3.10}$$

then, the attackers cannot decide whether the path is wave-pipelining or not by physical reverse engineering. Considering that the number of such paths can be enormous even for benchmarks of moderate size, and that it is usually expensive to determine the timing scheme for each path, the timing-based camouflaging strategy becomes very promising.

### 3.4.2 A Motivating Example

Before we formally describe our de-camouflaging framework, in this section, we introduce our TU design and use a motivating example to illustrate how the TUs can be used to recover the path timing schemes.

We follow the most widely used attack model [6, 7, 20, 27, 32, 45] and assume the attackers to have access to the camouflaged netlist and a functional circuit. The camouflaged netlist is acquired from physical reverse engineering. The timing for each path in the camouflaged netlist cannot be exactly determined with an inaccuracy factor $\tau$. The functional circuit is obtained from the open market with its operating clock period, i.e., $T_{clk}$, known by the attackers. The attackers can query the functional circuit as a black box to determine the correct outputs for the selected input vectors. The goal of the attackers is to determine the path timing based on the input–output pairs and recover the correct circuit functionality.

Now we use the example in Fig. 3.20 to introduce our TU design and illustrate our attack process. Consider the combinational path from $F_1$ to $F_3$ in the camouflaged netlist in Fig. 3.20b. Assume the path delay satisfies Eq. (3.10). To determine whether the path is a single-cycle path or a wave-pipelining path, we propose a novel TU design and insert it into the netlist as shown in Fig. 3.21a. Our TU design mainly consists of a flip flop and a multiplexer, which is controlled by a one-bit key. We denote the netlist after the insertion of TU as the transformed netlist. In the transformed netlist, we regard all the paths, including the combinational path from $F_1$ to $F_3$, the path from $F_1$ to $F_2'$, the path from $F_2'$ to $F_3$, etc., as single-cycle paths. Then, if the key bit is 0, the functionality of the transformed netlist is equivalent to that of the camouflaged netlist when the original path is a single-cycle

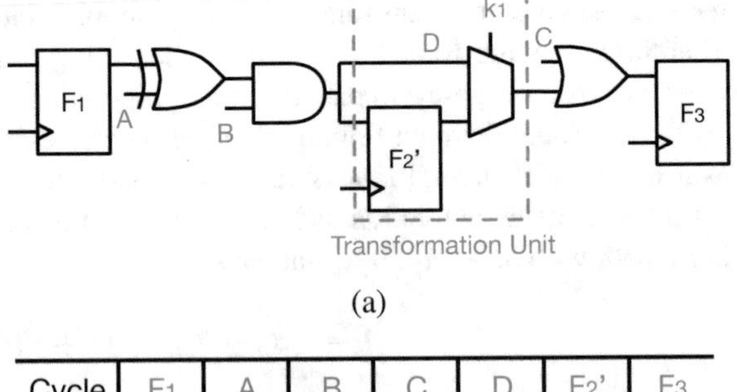


(a)

| Cycle | F1 | A | B | C | D | F2' | F3 |
|---|---|---|---|---|---|---|---|
| 0 | 1 | 0 | 1 | x | 1 | x | x |
| T | 1 | 0 | 0 | 0 | 0 | 1 | ? |

Observation: at $2T_{clk}$, if ? = 1, $k_1$=1, otherwise, $k_1$=0

(b)

**Fig. 3.21** An illustrative example to transform the camouflaged netlist with the proposed TU design and resolve the correct functionality based on input queries: (**a**) circuit transformation based on the proposed TU design; and (**b**) input queries to determine whether a path is single-cycle or wave-pipelining

path. Otherwise, the functionality of the transformed netlist is equivalent to that of the camouflaged netlist when the original path is a wave-pipelining path. Therefore, to determine whether the original path is single-cycle or wave-pipelining, it suffices to determine the key bit in the TU.

To determine the key bit, we need to query the black-box circuit. Since we assume full access to the scan chain, the logic values of $F_1$ and $F_3$ can be controlled and observed. For $F_2'$, as it does not exist in the real circuit, we cannot control/observe its logic value. Consider the input queries shown in Fig. 3.21b. At 0th cycle, the inputs $A$, $B$, $C$, and $F_1$ are set to 0, 1, $x$, 1, respectively, where $x$ represents don't care. Hence, $F_2'$ latches 1 at the clock rising edge at time $T_{clk}$. Then, we change the logic value for $A$, $B$, $C$, and $F_1$ to be 0, 0, 0, and 1, respectively. Now, the logic value of $D$ becomes 0, which is different from that of $F_2'$. Depending on the logic value latched by $F_3$ at time $2T_{clk}$, we can determine the key bit. More specifically, if $F_3$ latches 0, then, the key bit equals to 0; otherwise, the key bit equals to 1. Thereby, we can determine whether the path is wave-pipelining or not.

From the example described above, we see that it is indeed possible to determine whether a path is wave-pipelining or not by querying the black-box circuit and observing the outputs. To apply the attack procedure to a general circuit, there are following important questions we need to answer:

- How to insert the TUs to convert the camouflaged netlist to its single-cycle counterpart and guarantee the correct functionality can be achieved when proper keys are inserted;
- How to determine the input queries to prune incorrect keys in the presence of uncontrollable flip flops in the inserted TUs and prove the correctness of the resolved key bits;
- How to enhance the attack efficiency.

In the next section, we will describe our de-camouflaging framework with a focus on answering the questions above.

### 3.4.3 TimingSAT Framework

In this section, we formally describe the proposed TimingSAT framework. As shown in Fig. 3.22, TimingSAT consists of four steps. First, TUs are inserted into the camouflaged netlist to convert all the paths into single-cycle paths. Then, we unroll the transformed netlist and change all the uncontrollable flip flops into the circuit internal nodes. Although the SAT attack can be applied to the unrolled netlist directly, we propose a simplification procedure to reduce the size of the unrolled netlist and then exploit the SAT attack to prune the incorrect key assignments. Finally, a circuit post-processing step is conducted to reconstruct the de-camouflaged netlist.

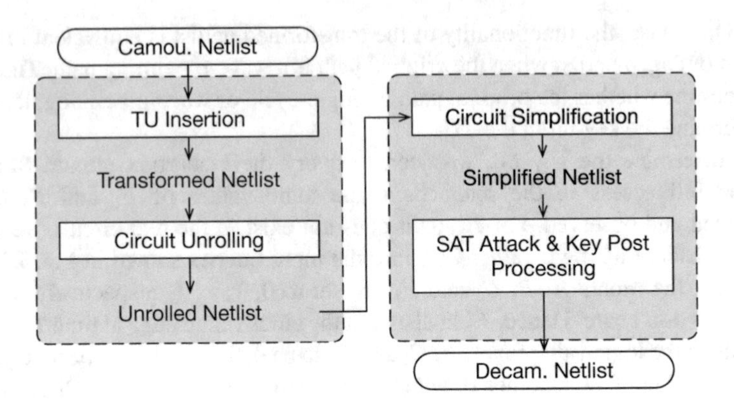


**Fig. 3.22** Flow chart of TimingSAT

## TU Insertion

Let $\mathscr{C}$ represent the black-box functional circuit and $C$ denote the camouflaged netlist extracted from physical reverse engineering. Since the flip flops in $C$ are controllable and observable through scan chain [20, 27, 32, 45], we regard the inputs and outputs of the flip flops the same as the primary outputs (POs) and primary inputs (PIs), respectively. Let $m$ be the total number of PIs with input space $\mathscr{I} \subseteq \{0, 1\}^m$ and $n$ be the total number of POs with output space $\mathscr{O} \subseteq \{0, 1\}^n$.

The first step of TimingSAT is to insert the TUs into $C$. In $C$, the wave-pipelining paths are generated by removing a subset of flip flops. Let d denote the subset of nodes in $C$ where flip flops are removed. To convert all the paths into single-cycle paths and guarantee that the correct functionality can be recovered when a proper set of keys are inserted, at least one TU needs to be inserted to each node in d. As d is unknown during the de-camouflaging process, a straightforward strategy is to insert a TU to each circuit node in $C$. However, such strategy can result in a huge amount of key bits, and, thus, a large number of possible functionalities for the camouflaged netlist, which can significantly increase the attack complexity [20].

We propose to leverage the estimated path delay to reduce the number of TUs inserted to $C$. Recall in $C$, there are two types of paths, i.e., single-cycle paths and wave-pipelining paths. For a single-cycle path in $C$, when a flip flop is inserted to the path, its functionality gets changed. This indicates for a TU inserted along a single-cycle path, its controlling key bit must be 0. Hence, all the TUs inserted along the single-cycle paths can be avoided. Considering the inaccuracy factor $\tau$ for the estimated path delay and the cycle period $T_{clk}$, we now have the following lemma.

**Lemma 3.2** *Consider a circuit node $u$ in $C$. If there exists a path $p$ through $u$ with delay $d_p$ that satisfies*

$$(1 + \tau)d_p < T_{clk},$$

**Algorithm 5** TU insertion

1: $V_t \leftarrow$ `TopologicalSort`$(C)$;
2: **for** $i = 1$ to $|V_t|$ **do**
3: $\quad u_i.\mathrm{dtopi} \leftarrow d_g + \min\{u_j.\mathrm{dtopi} \,|u_j \in u_i.\mathrm{fanin}\}$;
4: **end for**
5: **for** $i = |V_t|$ to 1 **do**
6: $\quad u_i.\mathrm{dtopo} \leftarrow d_g + \min\{u_j.\mathrm{dtopo} \,|u_j \in u_i.\mathrm{fanout}\}$;
7: **end for**
8: **for** $i = 1$ to $|V_t|$ **do**
9: $\quad$ **if** $u_i.\mathrm{dtopi} + u_i.\mathrm{dtopo} \geq T/(1+\tau)$ **then**
10: $\quad\quad$ `InsertTU`$(u_i)$;
11: $\quad$ **end if**
12: **end for**

*then, p must be a single-cycle path and no TUs need to be inserted at u.*

Based on Lemma 3.2, for each node in $C$, to decide whether a TU should be inserted, the key step is to determine the smallest delay for all the paths through the node. This can be finished in linear time with respect to the size of the camouflaged netlist. As shown in Algorithm 5, after the topological sort of the circuit nodes (line 1), we first traverse the circuit in a topological order and keep a record of the minimum delay from the PIs to each node (lines 2–4). Then, we traverse the circuit in a reverse topological order and determine the minimum delay from each node to the POs (line 5–7). The TUs can then be inserted based on the acquired delay and Lemma 3.2 (line 8–12).

### Determine Input Query Through Unrolling

Let $S$ denote the transformed netlist with $l$ TUs inserted. Then, the number of key bits in $S$ is $l$ with key space $\mathscr{K} \subseteq \{0, 1\}^l$. For $k \in \mathscr{K}$, let $S(k)$ denote the completed circuit by applying $k$ to $S$. All the paths in $S(k)$ are regarded as single-cycle paths. Consider a sequence of input vectors $I = \{i^0, \ldots, i^{p-1}\}$ of length $p$ with $i^j \in \mathscr{I}, \forall 0 \leq j \leq p-1$. Let $\mathscr{C}(I, s^0)$ denote the sequence of outputs produced by applying $I$ to $\mathscr{C}$ starting from an initial state $s^0$. We use $\mathscr{C}^j(I, s^0)$ to represent the output at $j$th cycle. Let $S(I, s^0; k)$ denote the sequence of outputs produced by $S(k)$ with the input sequence $I$ and the initial state $s^0$.

After the first step of TimingSAT, we can guarantee a TU is inserted to each node in d. Therefore, there must exist a key assignment $k^*$ such that $S(k^*)$ has the correct functionality. More formally, we have the following lemma.

**Lemma 3.3** *$\exists k^* \in \mathscr{K}$ such that $S(I, s^0; k^*) = \mathscr{C}(I, s^0)$, for any input sequence $I = \{i^0, \ldots, i^{p-1}\}$ of length $p > 0$ with $i^j \in \mathscr{I}, \forall 0 \leq j \leq p-1$, and initial state $s^0$.*

To determine $k^*$, [12, 34] propose the SAT attack to search for the input vectors that can be used to prune incorrect key assignments iteratively. However, these

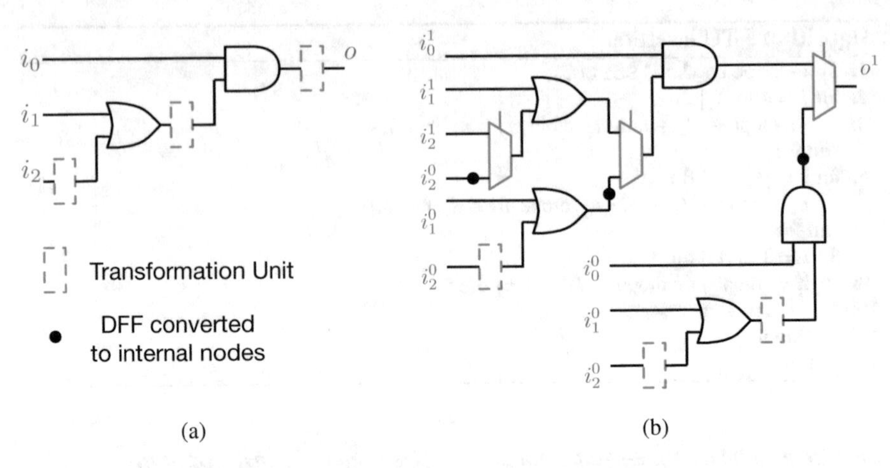


**Fig. 3.23** Example of circuit unrolling: (**a**) the transformed netlist; (**b**) the unrolled netlist generated by unrolling the transformed netlist for 1 time frame ($i_0^1$ represents the first input signal in the second time frame)

methods are mainly designed for combinational circuits.[1] In $S$, as the flip flops in the inserted TUs cannot be directly controlled or observed, existing SAT attack strategies cannot be applied directly.

To leverage the SAT attack, we propose to unroll $S$ to create a combinational logic that is equivalent to $S$ up to the unrolled time frame. In the unrolling process, we remove the uncontrollable flip flops in the TUs and connect their output signals to their input signals in the previous cycle. Meanwhile, different from the traditional circuit unrolling scheme [24], we only keep the POs in the last time frame. In Fig. 3.23, we show an example of the transformed netlist and the netlist generated by unrolling the transformed netlist for 1 time frame. As we can see, the unrolled netlist in Fig. 3.23b consists of the PIs for 2 time frames, e.g., $i_0^0$ and $i_0^1$, and the POs for the last time frame, i.e., $o^1$.

In Fig. 3.23b, as we unroll the netlist for 1 time frame, not all the flip flops in the inserted TUs are converted into circuit internal nodes. This indicates the outputs of the unrolled circuit still depend on the uncontrollable flip flops. In fact, for general sequential circuits, despite the number of unrolled time frames, the outputs of the generated netlist can always depend on the uncontrollable flip flops. This is because of the existence of the feedback arcs in the fanin cone of the uncontrollable flip flops. One example is shown in Fig. 3.24a. If $F_1$ is an uncontrollable flip flop, then, due to the feedback arc, the outputs of the unrolled netlist always depend on $F_1$.

However, in the transformed netlist $S$, there are no such feedback arcs. The reasons come from twofolds. On the one hand, after we regard the outputs and inputs

[1]Although SAT attack for sequential circuits has been proposed recently [11], it requires unbounded model checking to formally guarantee the correctness of the resolved netlist, which can be intractable.

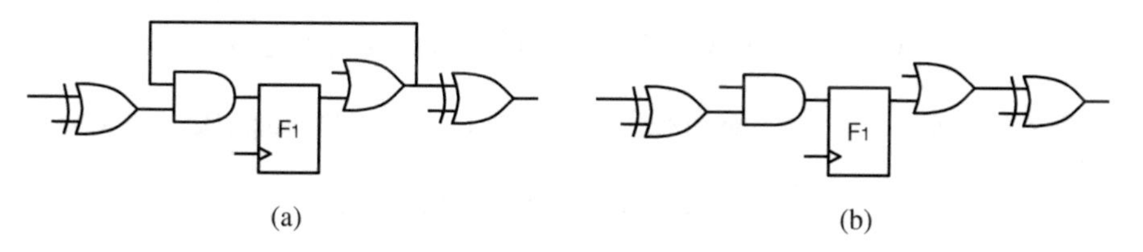


**Fig. 3.24** Example of circuits (**a**) with and (**b**) without feedback arcs. For (**a**), unrolling is insufficient to convert the flip flops into internal nodes

of controllable flip flops as PIs and POs, there are no feedback arcs in $C$.[2] On the other hand, the insertion of TUs does not introduce any feedback connections either. Therefore, let $d$ denote the maximum number of TUs along a path in $S$, we have the following lemma regarding the unrolled netlist.

**Lemma 3.4** *After unrolling $S$ for $d$ time frames, all the uncontrollable flip flops in $S$ can be converted into circuit internal nodes and the outputs of the unrolled netlist become independent of the uncontrollable flip flops.*

Let $R$ denote the netlist generated by unrolling $S$ for $d$ time frames. Then, the inputs of $R$ are composed of the PIs of $S$ for $d + 1$ cycles. For example, for the transformed netlist in Fig. 3.23a, $d = 3$. The netlist generated by unrolling the transformed netlist for 3 cycles is shown in Fig. 3.26a. As we can see, all the flip flops are converted into circuit internal nodes.

As the functionality of the circuit is preserved during unrolling, for any input sequence $I = \{i^0, \ldots, i^d\}$ of length $d + 1$ with $i^j \in \mathcal{I}, \forall 0 \le j \le d$, we have

$$S^d(I, s^0; k) = R(I; k). \tag{3.11}$$

As the output of $R$ is independent of initial state $s^0$, Eq. (3.11) holds regardless of the initial state $s^0$. Hence, we rewrite Eq. (3.11) as $S^d(I; k) = R(I; k)$.

After the circuit unrolling, as all the paths in $R$ operate within a single clock cycle and the outputs of $R$ are completely determined by the controllable inputs, we can now leverage the SAT attack proposed in [34] to search for $k^+ \in \mathcal{K}$ such that for any input sequence $I = \{i^0, \ldots, i^d\}$ of length $d + 1$ with $i^j \in \mathcal{I}, \forall 0 \le j \le d$,

$$R(I; k^+) = R(I; k^*). \tag{3.12}$$

Regarding $k^+$, we have the following theorem.

**Theorem 3.2** *$\forall k^+ \in \mathcal{K}$ that satisfies Eq.* (3.12), *given an initial state $s^0$ and input sequence $\{i^0, \ldots, i^p\}$ of length $p + 1$, where $p \ge d$, with $i^j \in \mathcal{I}, \forall 0 \le j \le p$, we have*

[2]We assume there are no combinational loops in the original netlist, which holds for most of the circuits.

$$S^j(i^0, \ldots, i^p, s^0; k^+) = \mathscr{C}^j(i^0, \ldots, i^p, s^0), \forall j \geq d.$$

***Proof 3.3*** For any $j \geq d$, based on Eqs. (3.11), (3.12), and Lemma 3.3, we have

$$\begin{aligned}
S^j(i^0, \ldots, i^p, s^0; k^+) &= S^j(i^{j-d}, \ldots, i^j; k^+) \\
&= R(i^{j-d}, \ldots, i^j; k^+) \\
&= R(i^{j-d}, \ldots, i^j; k^*) \\
&= S^j(i^{j-d}, \ldots, i^j; k^*) \\
&= S^j(i^0, \ldots, i^p, s^0; k^*) \\
&= \mathscr{C}^j(i^0, \ldots, i^p, s^0).
\end{aligned}$$

Hence proved.

Theorem 3.2 indicates that starting from a random initial state, after feeding the de-camouflaged netlist with $d + 1$ input vectors, we can always guarantee the outputs generated by the downstream execution of the circuit to be correct. Hence, we formally prove the correctness of the resolved circuit functionality.

#### Netlist Simplification for De-camouflaging Acceleration

Based on the unrolling scheme proposed above, we are able to leverage the existing SAT attack and find $k^+$. However, as the size of the unrolled netlist increases in proportional to $d$, the performance of the SAT attack can be significantly degraded for a circuit with large $d$. From our empirical study, which will be shown in the experimental results, for small benchmarks, when the number of unrolling cycles increases from 2 to 10, the runtime of the SAT attack on average increases by 45×. For large benchmark circuits, as shown in Fig. 3.25, $d$ can be as large as 82. Simply unrolling the circuit for 82 cycles will result in a much larger netlist compared to the original netlist, which makes the SAT attack prohibitively expensive and unpractical.

To enhance the efficiency of TimingSAT, we hope to reduce the number of unrolling time frames. According to [50], in order to satisfy the delay constraints in Eq. (3.9), for each wave-pipelining path generated by removing a flip flop, the flip flops at the beginning and the end of the path must remain unchanged. Therefore, to recover the original circuit, for each path in $C$, at most one flip flop needs to be inserted. Let $\mathscr{P}_R$ represent the set of paths in $R$ and $\mathscr{K}_p$ represent the set of key bits along a path $p \in \mathscr{P}_R$, then, we have the following lemma.

**Lemma 3.5** $\sum_{j \in \mathscr{K}_p} k_j^* \leq 1, \forall p \in \mathscr{P}_R$.

Based on Lemma 3.5, we can identify the redundant paths in $R$ and try to simplify $R$. Again consider the camouflaged netlist in Fig. 3.23a and the unrolled netlist in

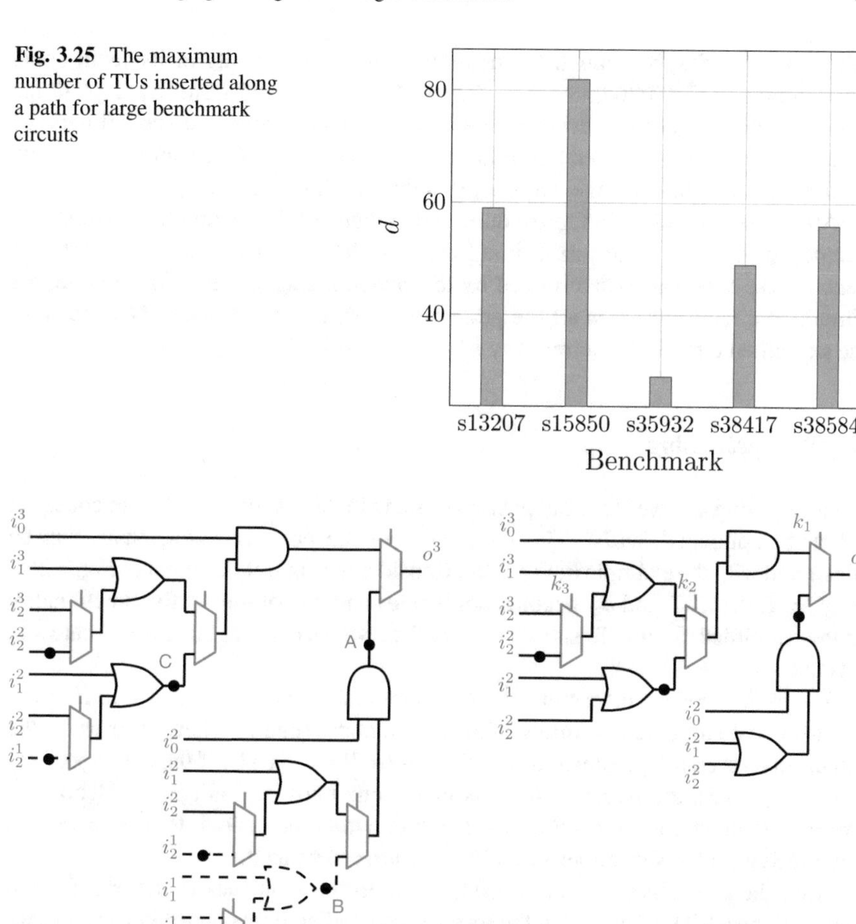


**Fig. 3.25** The maximum number of TUs inserted along a path for large benchmark circuits

**Fig. 3.26** Example of (**a**) the unrolled netlist with only controllable PIs as circuit inputs for the camouflaged netlist in Fig. 3.23a and (**b**) the simplified netlist generated by removing the redundant paths

Fig. 3.26a. In Fig. 3.26a, the paths marked with dashed lines are redundant as they all have more than one non-zero key bits along the paths. Therefore, we can remove the redundant paths and get the simplified netlist as shown in Fig. 3.26b.

The simplification procedure can be accomplished in linear time with respect to the circuit size by traversing the unrolled circuit $R$ in a reverse topological order. Starting from the POs, for each node converted from a flip flop in the unrolling process, we check whether there are any other nodes converted from flip flops in its fanout cone. If there are no such nodes in the fanout cone, this node is kept;

otherwise, we simply remove the node as well as the circuits in its fanin cone. For example in Fig. 3.26a, nodes $A$, $B$, and $C$ are converted from flip flops in the unrolling process. For $B$, because there is $A$ in its fanout cone, we know it must be redundant, and, thus, we remove it and its fanin cone. For $C$, because there is no such nodes in its fanout cone, we keep it in the simplified netlist.

After the simplification procedure, the inputs of the simplified netlist are composed of PIs of $S$ for just 2 time frames, which remains constant for different benchmarks and is only determined by the camouflaging strategy. Meanwhile, the simplified circuit preserves all the properties of $R$. By applying the SAT attack to the simplified circuit, the correct key $k^+$ can be resolved.

### Key Post-processing

After resolving $k^+$, we carry out post-processing to the resolved key bits to construct the de-camouflaged netlist. The importance of the post-processing stage can be illustrated with the example in Fig. 3.26. Consider the simplified netlist in Fig. 3.26b. If $k_1^+ = 1$, then, $k_2^+$ and $k_3^+$ become don't cares and do not impact the functionality of the simplified circuit. To satisfy Lemma 3.5, we need set all these don't-care key bits to 0.

The pseudo code to determine the key values is shown in Algorithm 6. In $R$, given $k^+$, we first do topological sorting of the circuit nodes (line 1). Then, we traverse the circuit in a reverse topological order (lines 2–6). For each TU, if the corresponding key is 1 (line 3), all the key bits in its fanin cone must be don't cares. Hence, we traverse its fanin cone and set all the key bits within the cone to 0. The process is continued until the key values of all the TUs are determined.

After the key values are determined, we insert the keys back to the transformed netlist. For each TUs inserted in the transformed netlist, if its control key bit is 0, we just remove the TU and connect the input of the TU directly to its output. Otherwise, we remove the TU and insert a flip flop to the circuit node. Thereby, we can construct a sequential circuit which operates within a single clock period and has the correct functionality.

**Algorithm 6** Post-processing to determine the key values

1: $V_t \leftarrow$ `TopologicalSort`$(R)$;
2: **for** $i = |V_t|$ to 1 **do**
3:     **if** $u_i$.gate = TU and $k_i^+ = 1$ **then**
4:         `SetKeyInFanin`$(u_i, R)$
5:     **end if**
6: **end for**

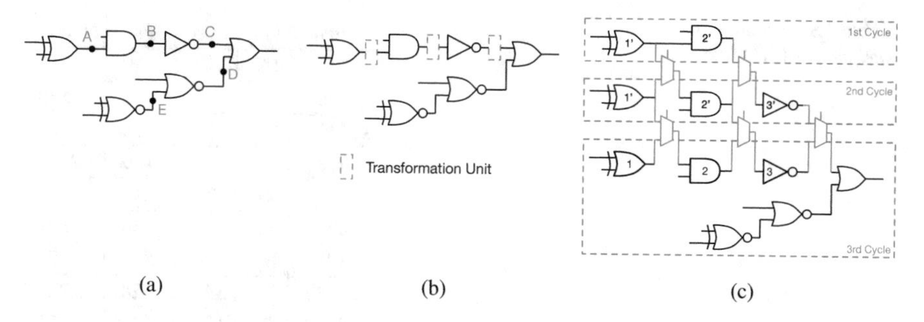


**Fig. 3.27** Example of the transformation flow to enable SAT attack when there exist at most three waves for a wave-pipelining path: (**a**) camouflaged netlist; (**b**) insertion of TUs; and (**c**) simplified unrolled circuit

## Discussion

In TimingSAT, we rely on Lemma 3.5 to simplify the unrolled circuit and enhance the efficiency of TimingSAT significantly. As we have discussed, Lemma 3.5 holds since the existing camouflaged netlists only consist of wave-pipelining paths with at most two logic waves. However, it is indeed possible to remove multiple flip flops and generate wave-pipelining paths with more than two logic waves while keeping the circuit functionality unchanged [49]. Such changes can be easily accommodated in our current framework. First, from physical reverse engineering, though accurate path delay cannot be determined due to the uncertainty, the upper bound on the path delay can still be estimated to determine the maximum number of logic waves along a path. Then, in the simplification process, for each node converted from a flip flop, in its fanout cone, we can determine the maximum number of nodes converted from flip flops along a certain path. If the number is larger than the maximum number of logic waves, we know the path must be redundant. Hence, we can remove the node as well as its fanin cone. One example of the flow is shown in Fig. 3.27. Assume there are at most 3 logic waves for a wave-pipelining paths. Given the camouflaged netlist in Fig. 3.27a, we first determine whether a TU should be inserted to each node, i.e., $A$, $B$, $C$, $D$, $E$. Assume nodes $D$ and $E$ satisfy Lemma 3.2, then, we only insert TUs to nodes $A$, $B$, and $C$ and get the transformed netlist shown in Fig. 3.27b. After we unroll and simplify the transformed netlist, we get the simplified netlist in Fig. 3.27c. Finally, we apply the SAT attack to the simplified netlist and carry out post-processing to determine the correct key bits.

### *3.4.4 Experimental Results*

In this section, we report on our extensive experiments to demonstrate the effectiveness of the proposed TimingSAT framework. The transformation and simplification procedure in TimingSAT is implemented in C++ and the SAT attack engine is

**Table 3.9** Benchmark statistics of ISCAS'98

| Bench | PIs | POs | FFs | Nodes |
|---|---|---|---|---|
| s953 | 16 | 23 | 29 | 418 |
| s1196 | 14 | 14 | 18 | 530 |
| s1238 | 14 | 14 | 18 | 509 |
| s5378 | 35 | 49 | 179 | 2779 |
| s9234 | 36 | 39 | 211 | 5597 |
| s13207 | 62 | 152 | 638 | 8027 |
| s15850 | 77 | 150 | 534 | 9786 |
| s35932 | 35 | 320 | 1728 | 16,353 |
| s38417 | 28 | 106 | 1636 | 22,397 |
| s38584 | 38 | 304 | 1426 | 19,407 |

adopted from [34]. The timing-based camouflaging strategies are implemented following [50] and the functional camouflaging strategies are implemented following [27]. The benchmarks are chosen from ISCAS'89 benchmark suite [4], the detailed statistics of which are shown in Table 3.9. We run all the experiments on an eight-core 3.40 GHz Linux server with 32 GB RAM. We set the runtime limit of TimingSAT to $5 \times 10^4$ s.

#### Efficiency of TimingSAT

We first report the efficiency of TimingSAT on the benchmark circuits and demonstrate the importance of the simplification procedure. To generate the camouflaged netlist, 10 flip flops are removed following [50]. We also assume the largest uncertainty in the reverse engineering process, i.e., we assume the delay of each path in the camouflaged netlist satisfies Eq. (3.10) so that a TU is inserted to each circuit node in the transformation process. In Table 3.10, we show the statistics of the unrolled netlists without simplification and the simplified netlists. As we can see, the simplified netlists are significantly smaller compared with the unrolled netlist. Then we conduct SAT attacks on the unrolled netlist and the simplified netlist. As shown in Table 3.10, for the first three small benchmarks, TimingSAT with the simplification procedure achieves on average 87× speedup. Meanwhile, for the large benchmarks, when we directly apply the SAT attacks to the unrolled netlists, the SAT attacks cannot be finished within the pre-defined time limit. However, with the simplification procedure, all the camouflaged netlist can be de-camouflaged efficiently within $10^4$ s, which demonstrates the importance of the simplification procedures and the efficiency of our TimingSAT framework.

#### Runtime Dependency of TimingSAT

Now, we want to evaluate the runtime dependency of TimingSAT on the number of flip flops removed in the camouflaging process and the delay uncertainty in

**Table 3.10** Performance of TimingSAT with and without simplification on the benchmark circuits

| | TimingSAT w/o simplification | | | | | | TimingSAT w/simplification | | | | | |
|---|---|---|---|---|---|---|---|---|---|---|---|---|
| Bench | PIs | POs | Nodes | Keys | # Iter | Rt (s) | PIs | POs | Nodes | Keys | # Iter | Rt (s) |
| s953 | $5.6\times10^2$ | 42 | $6.6\times10^3$ | $1.8\times10^2$ | $1.2\times10^2$ | 36.5 | 70 | 42 | $1.0\times10^3$ | $1.8\times10^2$ | 72 | 1.7 |
| s1196 | $5.0\times10^2$ | 22 | $1.2\times10^4$ | $5.1\times10^2$ | $5.4\times10^2$ | $8.6\times10^2$ | 44 | 22 | $1.6\times10^3$ | $5.1\times10^2$ | $1.0\times10^2$ | 4.5 |
| s1238 | $4.6\times10^3$ | 22 | $1.1\times10^4$ | $4.9\times10^2$ | $2.7\times10^2$ | $2.6\times10^2$ | 44 | 22 | $1.5\times10^3$ | $4.9\times10^2$ | $1.2\times10^2$ | 5.4 |
| s5378 | $5.2\times10^3$ | $2.0\times10^2$ | $7.4\times10^4$ | $2.5\times10^3$ | N/A | N/A | $3.8\times10^2$ | $2.0\times10^2$ | $8.4\times10^3$ | $2.5\times10^3$ | $2.5\times10^2$ | 82.1 |
| s9234 | $1.3\times10^4$ | $2.4\times10^2$ | $3.1\times10^5$ | $5.3\times10^3$ | N/A | N/A | $4.7\times10^2$ | $2.4\times10^2$ | $1.6\times10^4$ | $5.3\times10^3$ | $1.3\times10^3$ | $1.6\times10^3$ |
| s13207 | $4.0\times10^4$ | $7.8\times10^2$ | $4.3\times10^5$ | $7.2\times10^3$ | N/A | N/A | $1.3\times10^3$ | $7.8\times10^2$ | $2.3\times10^4$ | $7.2\times10^3$ | $6.3\times10^2$ | $4.7\times10^2$ |
| s15850 | $4.8\times10^4$ | $6.7\times10^2$ | $7.5\times10^5$ | $1.0\times10^4$ | N/A | N/A | $1.2\times10^3$ | $6.7\times10^2$ | $2.9\times10^4$ | $1.0\times10^4$ | $1.1\times10^3$ | $1.3\times10^3$ |
| s35932 | $4.9\times10^4$ | $2.0\times10^3$ | $4.2\times10^5$ | $1.8\times10^4$ | N/A | N/A | $3.5\times10^3$ | $2.0\times10^3$ | $4.8\times10^4$ | $1.8\times10^4$ | $1.2\times10^2$ | $4.4\times10^2$ |
| s38417 | $7.9\times10^4$ | $1.7\times10^3$ | $1.0\times10^6$ | $2.0\times10^4$ | N/A | N/A | $3.3\times10^3$ | $1.7\times10^3$ | $6.7\times10^4$ | $2.0\times10^4$ | $1.6\times10^3$ | $6.0\times10^3$ |
| s38584 | $7.9\times10^4$ | $1.7\times10^3$ | $1.0\times10^6$ | $1.7\times10^4$ | N/A | N/A | $2.9\times10^3$ | $1.7\times10^3$ | $5.7\times10^4$ | $1.7\times10^4$ | $5.5\times10^2$ | $2.5\times10^3$ |

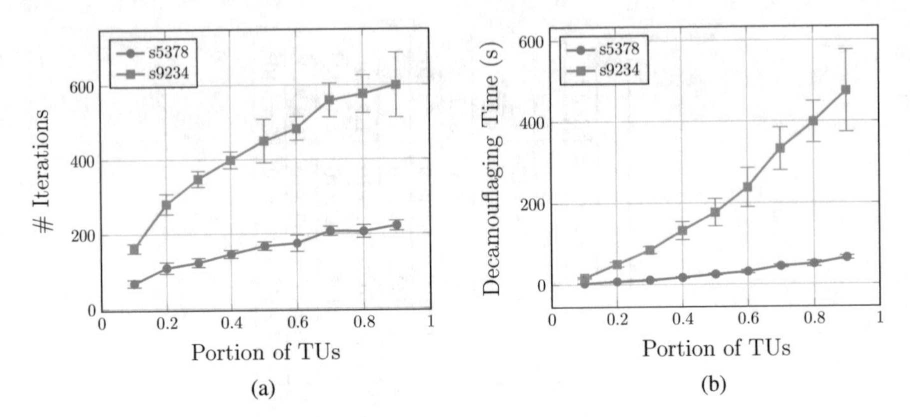


**Fig. 3.28** Impact of delay uncertainty on the de-camouflaging complexity. (**a**) Number of iterations. (**b**) De-camouflaging time

the reverse engineering process. We use benchmark circuits s5378 and s9234 as examples. To evaluate the impact of delay uncertainty, we fix the number of removed flip flops to 10 in the camouflaging process. Then, we gradually increase the number of TUs inserted in the transformation process and run the de-camouflaging attack. For each configuration, we run 10 different experiments and insert the TUs to different subsets of nodes for each experiment. We show the de-camouflaging iterations and time in Fig. 3.28. As we can see, with the increase of the number of inserted TUs, a significant increase of the de-camouflaging iterations and time can be observed. Specifically, when TUs are inserted to 90% of circuit nodes, the de-camouflaging time is almost 30× compared to the case when TUs are inserted to 10% of circuit nodes.

To evaluate the impact of the number of flip flops removed in the camouflaging process, we select the largest uncertainty and insert a TU to each node in the camouflaged netlist. Then, we gradually change the number of removed flip flops and run the de-camouflaging attack. We run 10 different experiments and remove different flip flops for each experiment. The change of de-camouflaging iterations and time is shown in Fig. 3.29. As can be observed, while the de-camouflaging time for s5378 increases slightly with the increase of the number of removed flip flops, the de-camouflaging time for s9234 even reduces. This indicates when the number of unrolling time frames and the number of inserted TUs remain unchanged, the de-camouflaging time is not directly dependent on the number of removed flip flops in the camouflaging process.

#### Impact of Unrolling Time Frames

As have described in Sect. 3.4.3, Lemma 3.5 is the key basis that enables the simplification process without impacting the correctness of the resolved keys.

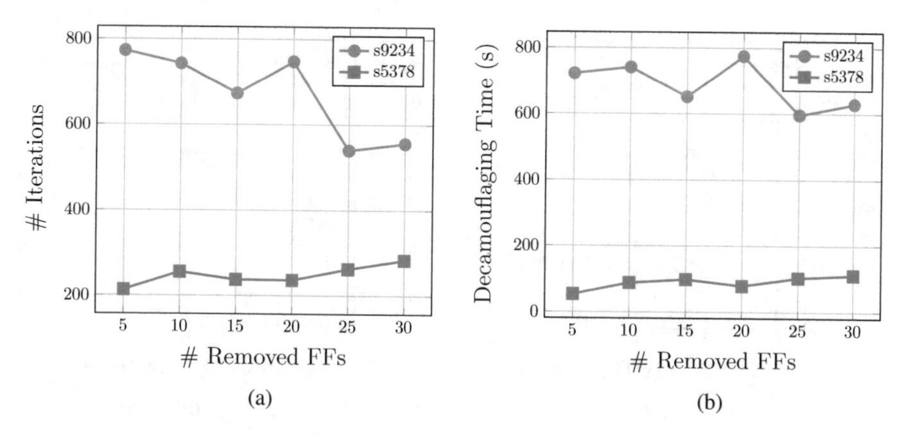


**Fig. 3.29** Impact of the number of removed flip flops on the de-camouflaging complexity. (**a**) Number of iterations. (**b**) De-camouflaging time

Meanwhile, Lemma 3.5 holds as the existing camouflaged netlists only consist of wave-pipelining paths with at most two logic waves. Now, we want to evaluate the impact when more than two logic waves exist in the wave-pipelining paths. We select three small benchmark circuits, including s953, s1196, and s1238, and two large benchmark circuits, i.e., s5378 and s13207, as examples and demonstrate the relation between the number of possible logic waves along a wave-pipelining paths and the de-camouflaging iterations and time. We still assume the largest uncertainty in the reverse engineering process and remove 10 flip flops following [50]. As shown in Figs. 3.30 and 3.31, both the de-camouflaging time and iterations increase significantly with the number of logic waves. Specifically, for the three small benchmarks, increasing the number of logic waves from 2 to 10 can on average increase the de-camouflaging time by 45.8×. For the two large benchmarks, when the number of logic waves becomes larger than 4, TimingSAT cannot finish within the pre-defined time limit. The experimental results indicate that increasing the number of logic waves for the wave-pipelining paths can indeed enhance the security against SAT attacks.

However, since we assume the largest uncertainty here, the result is rather conservative. We further investigate the change of de-camouflaging iterations and time when both the number of unrolling time frames and the number of inserted TUs are changed. The experimental results are shown in Fig. 3.31. As we can see, when we insert TUs to 50% or 20% of circuit nodes, significant reduction of de-camouflaging time can be observed. Even when the number of unrolling time frames becomes as large as 10, TimingSAT can still be finished within the pre-defined time limit.

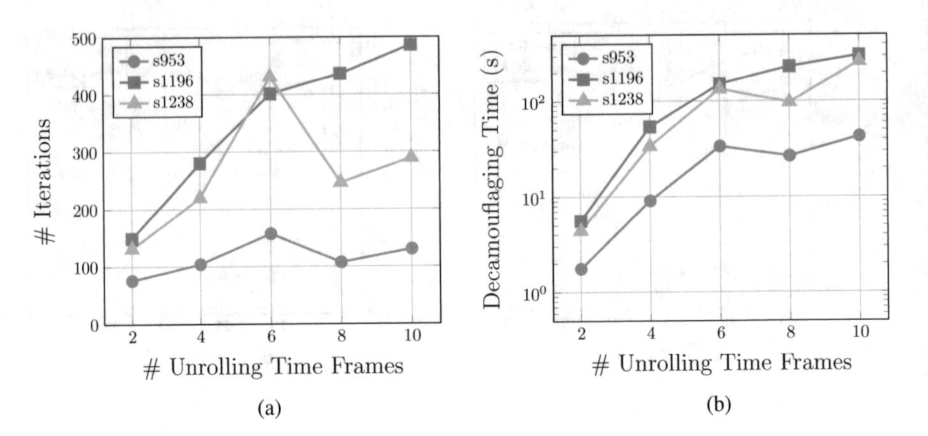


**Fig. 3.30** Relation between de-camouflaging complexity with the number of unrolling time frames. (**a**) Number of iterations. (**b**) De-camouflaging time

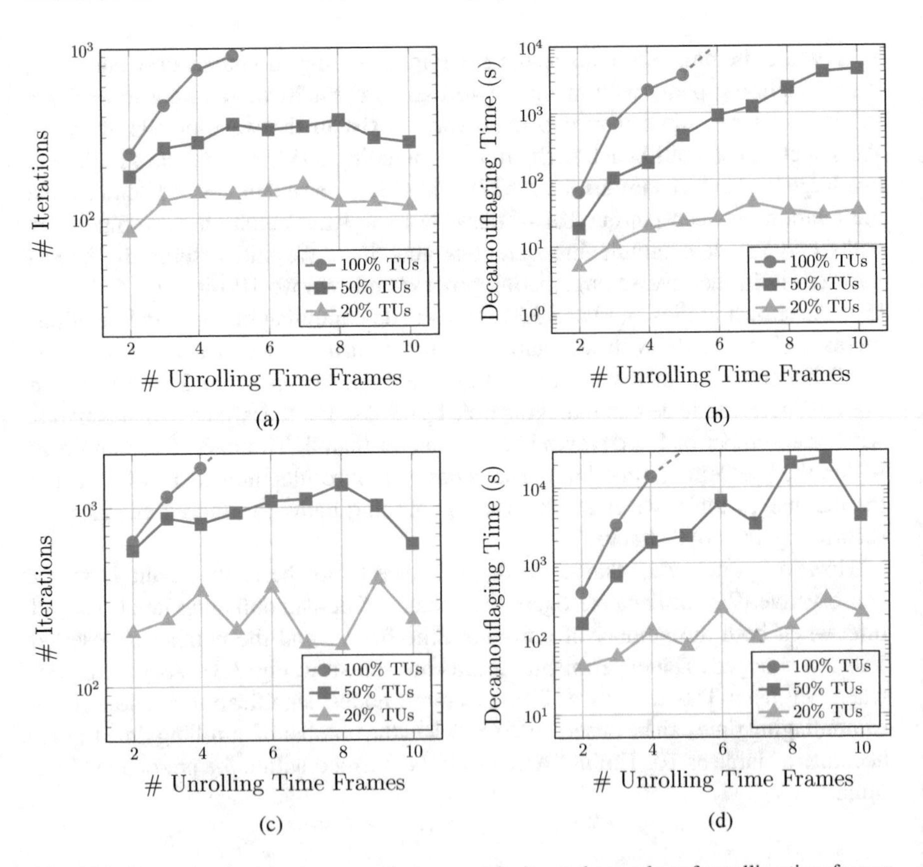


**Fig. 3.31** Dependency of the de-camouflaging complexity on the number of unrolling time frames and the number of inserted TUs. (**a**) Number of iterations (s5378). (**b**) De-camouflaging time (s5378). (**c**) Number of iterations (s13207). (**d**) De-camouflaging time (s13207)

### De-camouflaging of Combination of Timing-Based and Traditional Strategy

Now, we evaluate the situation when the timing-based camouflaging strategy is combined with high-entropy functional camouflaging strategies. We leverage the fault-analysis based camouflaging strategy proposed in [27] and combine it with the timing-based strategy. We use benchmarks s5378 and s9234 as examples. The number of removed flip flops is set to 10 and the largest delay uncertainty is assumed. We gradually change the number of camouflaging gates inserted into the circuit and compare the de-camouflaging time of the combined strategy with the camouflaging strategy in [27]. We run 10 experiments for each configuration and insert the camouflaging cells into different circuit nodes. The experimental results are shown in Fig. 3.32. As we can see, by combining different camouflaging strategies, much larger de-camouflaging time can be achieved compared with

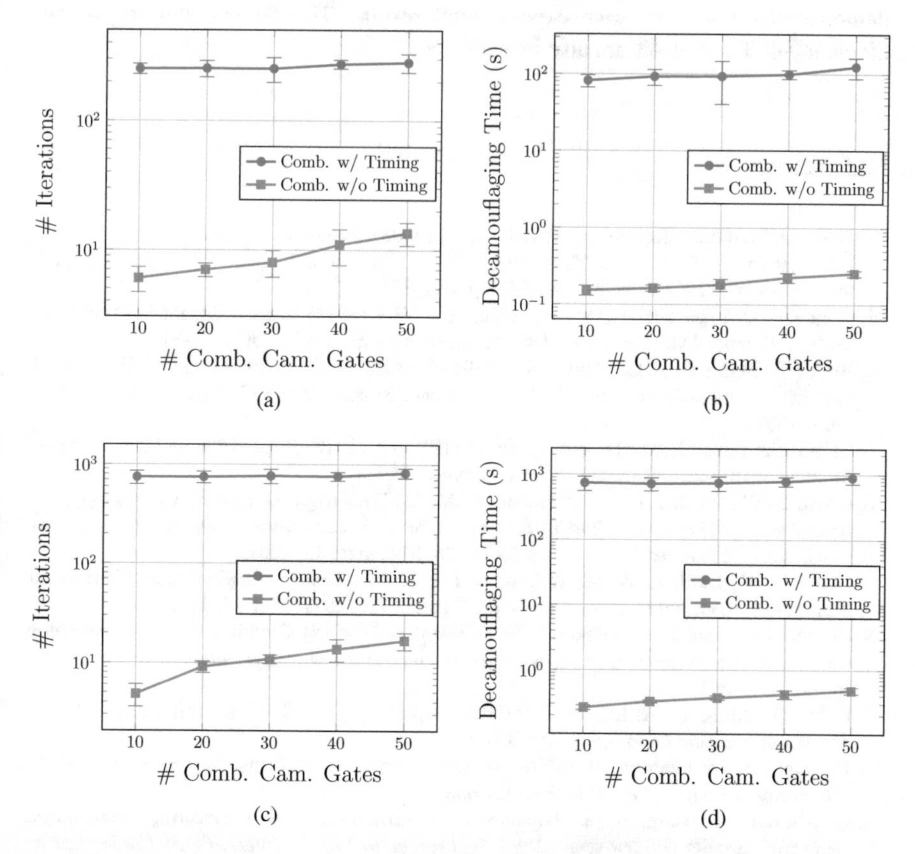


**Fig. 3.32** Comparison on the de-camouflaging complexity between random camouflaging with/without timing camouflaging. (**a**) Number of iterations (s5378). (**b**) De-camouflaging time (s5378). (**c**) Number of iterations (s9234). (**d**) De-camouflaging time (s9234)

leveraging the strategy in [27] alone. However, even after 50 camouflaging gates are inserted following [27], the netlist can still be de-camouflaged within $10^3$ s.

### 3.4.5 Summary

In Sect. 3.4, we propose TimingSAT, a SAT-based framework for security evaluation of the timing-based parametric camouflaging strategies. TimingSAT leverages a novel transformation strategy to convert all the paths in the camouflaged netlists into single-cycle paths while preserving the correct circuit functionalities. An unrolling procedure is then proposed in TimingSAT to enable SAT attacks. The functional correctness of the resolved netlist is formally proved. To accelerate TimingSAT, we also propose a simplification procedure to significantly reduce the complexity of the transformed netlist. The efficiency and effectiveness of TimingSAT are demonstrated with extensive experimental results. The factors that impact the efficiency of TimingSAT are also investigated.

## References

1. 16nm LP Predictive Technology Model ver. 2.1 (2008). http://ptm.asu.edu.
2. Baumgarten, A., Tyagi, A., & Zambreno, J. (2010). Preventing IC piracy using reconfigurable logic barriers. *IEEE Design & Test of Computers, 27*(1), 66–75.
3. Becker, G. T., Regazzoni, F., Paar, C., & Burleson, W. P. (2014). Stealthy dopant-level hardware Trojans: Extended version. *Journal of Cryptographic Engineering, 4*(1), 19–31.
4. Brglez, F., Bryan, D., & Kozminski, K. (1989). Combinational profiles of sequential benchmark circuits. In *Proceedings of the IEEE International Symposium on Circuits and Systems* (pp. 1929–1934).
5. Chipworks. Intel's 22-nm Tri-gate Transistors Exposed. (2012). http://www.eet.bme.hu/mizsei/Montech/intel-s-22-nm-trigate-transistors-exposed.html.
6. Chow, L.-W., Baukus, J. P., & Clark Jr, W. M. (2007). *Integrated circuits protected against reverse engineering and method for fabricating the same using an apparent metal contact line terminating on field oxide*. US Patent 7,294,935. November 13, 2007.
7. Chow, L.-W., Clark Jr, W. M., & Baukus, J. P. (2007). *Covert transformation of transistor properties as a circuit protection method*. US Patent 7,217,977. May 15, 2007.
8. Cocchi, R. P., Baukus, J. P., Chow, L. W., & Wang, B. J. (2014). Circuit camouflage integration for hardware IP protection. In *Proceedings of the IEEE/ACM Design Automation Conference* (pp. 153:1–153:5).
9. Cohn, D., Atlas, L., & Ladner, R. (1994). Improving generalization with active learning. *Journal of Machine Learning, 15*(2), 201–221.
10. Dasgupta, S., & Langford, J. (2009). A tutorial on active learning. In *Proceedings of the International Conference on Machine Learning*.
11. El Massad, M., Garg, S., & Tripunitara, M. (2017). Reverse engineering camouflaged sequential circuits without scan access. In *Proceedings of the International Conference on Computer Aided Design* (pp. 33–40). Piscataway: IEEE.
12. El Massad, M., Garg, S., & Tripunitara, M. V. (2015). Integrated circuit (IC) decamouflaging: Reverse engineering camouflaged ICs within minutes. In *Proceedings of the Network and Distributed System Security Symposium*.

13. Erbagci, B., Erbagci, C., Akkaya, N. E. C., & Mai, K. (2016). A secure camouflaged threshold voltage defined logic family. In *Proceedings of the IEEE International Symposium on Hardware Oriented Security and Trust* (pp. 229–235).
14. Hanneke, S. (2007). A bound on the label complexity of agnostic active learning. In *Proceedings of the International Conference on Machine Learning* (pp. 353–360).
15. Jin, Y. (2015). Introduction to hardware security. *Electronics, 4*(4), 763–784.
16. Kullback, S. (1968). *Information theory and statistics*. North Chelmsford: Courier Corporation.
17. Lee, Y. W., & Touba, N. A. (2015). Improving logic obfuscation via logic cone analysis. In *Proceedings of the IEEE Latin-American Test Symposium* (pp. 1–6).
18. Li, L., & Zhou, H. (2013). Structural transformation for best-possible obfuscation of sequential circuits. In *Proceedings of the IEEE International Symposium on Hardware Oriented Security and Trust* (pp. 55–60).
19. Li, M., Shamsi, K., Jin, Y., & Pan, D. Z. (2018). TimingSAT: Decamouflaging timing-based logic obfuscation. In *Proceedings of the IEEE International Test Conference*.
20. Li, M., Shamsi, K., Meade, T., Zhao, Z., Yu, B., Jin, Y., et al. (2016). Provably secure camouflaging strategy for IC protection. In *Proceedings of the International Conference on Computer Aided Design* (pp. 28:1–28:8).
21. Malik, S., Becker, G. T., Paar, C., & Burleson, W. P. (2015). Development of a layout-level hardware obfuscation tool. In *Proceedings of the IEEE Annual Symposium on VLSI* (pp. 204–209).
22. Meade, T., Zhao, Z., Zhang, S., Pan, D., & Jin, Y. (2017). Revisit sequential logic obfuscation: Attacks and defenses. In *Proceedings of the IEEE International Symposium on Circuits and Systems* (pp. 1–4). Piscataway: IEEE.
23. Mentor Graphics. (2008). Calibre verification user's manual.
24. Miskov-Zivanov, N., & Marculescu, D. (2007). Soft error rate analysis for sequential circuits. In *Proceedings of the Design, Automation and Test in Europe*. Piscataway: IEEE.
25. NanGate FreePDK45 Generic Open Cell Library. (2008). http://www.si2.org/openeda.si2.org/projects/nangatelib
26. Quadir, S. E., Chen, J., Forte, D., Asadizanjani, N., Shahbazmohamadi, S., Wang, L., et al. (2016). A survey on chip to system reverse engineering. *ACM Journal on Emerging Technologies in Computing Systems, 13*(1), 6:1–6:34.
27. Rajendran, J., Sam, M., Sinanoglu, O., & Karri, R. (2013). Security analysis of integrated circuit camouflaging. In *Proceedings of the ACM Conference on Computer & Communications Security* (pp. 709–720).
28. Rezaei, A., Shen, Y., Kong, S., Gu, J., & Zhou, H. (2018). Cyclic locking and memristor-based obfuscation against CycSAT and inside foundry attacks. In *Proceedings of the Design, Automation and Test in Europe* (pp. 85–90). Piscataway: IEEE.
29. Roy, J. A., Koushanfar, F., & Markov, I. L. (2008). EPIC: Ending piracy of integrated circuits. In *Proceedings of the Design, Automation and Test in Europe* (pp. 1069–1074).
30. Shamsi, K., Li, M., Meade, T., Zhao, Z., Pan, D. Z., & Jin, Y. (2017). AppSAT: Approximately deobfuscating integrated circuits. In *Proceedings of the IEEE International Symposium on Hardware Oriented Security and Trust*.
31. Shamsi, K., Li, M., Meade, T., Zhao, Z., Pan, D. Z., & Jin, Y. (2017). Circuit obfuscation and oracle-guided attacks: Who can prevail? In *Proceedings of the IEEE Great Lakes Symposium on VLSI*.
32. Shamsi, K., Li, M., Meade, T., Zhao, Z., Pan, D. Z., & Jin, Y. (2017). Cyclic obfuscation for creating sat-unresolvable circuits. In *Proceedings of the IEEE Great Lakes Symposium on VLSI*.
33. Shamsi, K., Li, M., Pan, D. Z., & Jin, Y. (2018). Cross-lock: Dense layout-level interconnect locking using cross-bar architectures. In *Proceedings of the IEEE Great Lakes Symposium on VLSI*.
34. Subramanyan, P., Ray, S., & Malik, S. (2015). Evaluating the security of logic encryption algorithms. In *Proceedings of the IEEE International Symposium on Hardware Oriented Security and Trust* (pp. 137–143).

35. Subramanyan, P., Tsiskaridze, N., Li, W., Gascon, A., Tan, W. Y., Tiwari, A., et al. (2014). Reverse engineering digital circuits using structural and functional analyses. *IEEE Transactions on Emerging Topics in Computing, 2*(1), 63–80.
36. Torrance, R., & James, D. (2011). The state-of-the-art in semiconductor reverse engineering. In *Proceedings of the IEEE/ACM Design Automation Conference* (pp. 333–338).
37. Vijayakumar, A., Patil, V. C., Holcomb, D. E., Paar, C., & Kundu, S. (2017). Physical design obfuscation of hardware: A comprehensive investigation of device and logic-level techniques. *IEEE Transactions on Information Forensics and Security, 12*(1), 64–77.
38. Wang, X., Zhou, Q., Cai, Y., & Qu, G. (2018). A conflict-free approach for parallelizing sat-based de-camouflaging attacks. In *Proceedings of the Asia and South Pacific Design Automation Conference*. Piscataway: IEEE.
39. Wiener, R. (2007). *An Algorithm for Learning Boolean Functions for Dynamic Power Reduction*. PhD Thesis, Department of Computer Science, University Of Haifa.
40. Xie,Y., & Srivastava, A. (2016). Mitigating SAT attack on logic locking. In *Proceedings of the International Conference on Cryptographic Hardware and Embedded Systems* (pp. 127–146).
41. Xie, Y., & Srivastava, A. (2017). Delay locking: Security enhancement of logic locking against IC counterfeiting and overproduction. In *Proceedings of the IEEE/ACM Design Automation Conference*.
42. Yang, S. (1991). Logic synthesis and optimization benchmarks user guide: Version 3.0. Technical report, MCNC Technical Report.
43. Yasin, M., Mazumdar, B., Rajendran, J., & Sinanoglu, O. (2016). SARLock: SAT attack resistant logic locking. In *Proceedings of the IEEE International Symposium on Hardware Oriented Security and Trust* (pp. 236–241).
44. Yasin, M., Mazumdar, B., Rajendran, J. J., & Sinanoglu, O. (2017). TTLock: Tenacious and traceless logic locking. In *Proceedings of the IEEE International Symposium on Hardware Oriented Security and Trust* (pp. 166–166). Piscataway: IEEE.
45. Yasin, M., Mazumdar, B., Sinanoglu, O., & Rajendran, J. (2016). CamoPerturb: Secure IC camouflaging for minterm protection. In *Proceedings of the International Conference on Computer Aided Design* (pp. 29:1–29:8).
46. Yasin, M., Mazumdar, B., Sinanoglu, O., & Rajendran, J. (2017). Security analysis of Anti-SAT. In *Proceedings of the Asia and South Pacific Design Automation Conference*.
47. Yasin, M., Sengupta, A., Nabeel, M. T., Ashraf, M., Rajendran, J. J., & Sinanoglu, O. (2017). Provably-secure logic locking: From theory to practice. In *Proceedings of the ACM Conference on Computer & Communications Security* (pp. 1601–1618). New York: ACM.
48. Yu, C., Zhang, X., Liu, D., Ciesielski, M., & Holcomb, D. (2017). Incremental SAT-based reverse engineering of camouflaged logic circuits. *IEEE Transactions on Computer-Aided Design of Integrated Circuits and Systems, 36*(99), 1–1.
49. Zhang, L., Li, B., Hashimoto, M., & Schlichtmann, U. (2018). Virtualsync: Timing optimization by synchronizing logic waves with sequential and combinational components as delay units. In *Proceedings of the IEEE/ACM Design Automation Conference*. Piscataway: IEEE.
50. Zhang, L., Li, B., Yu, B., Pan, D. Z., & Schlichtmann, U. (2018). TimingCamouflage: Improving circuit security against counterfeiting by unconventional timing. In *Proceedings of the Design, Automation and Test in Europe*.
51. Zhou, H., Jiang, R., & Kong, S. (2017). CycSAT: SAT-based attack on cyclic logic encryptions. In *Proceedings of the International Conference on Computer Aided Design* (pp. 49–56). Piscataway: IEEE.

# Chapter 4
# Fault Attack Protection and Evaluation

## 4.1 Introduction

The previous chapter has introduced our IC camouflaging optimization and evaluation algorithms to prevent reverse engineering. This chapter focuses on protecting hardware IP against fault injection attacks [3, 24, 49]. Different from passive side channel analysis [2, 18], fault attack actively injects errors into hardware, which can lead to leakage of critical information [5, 17, 45] and nullification of security policies [46, 48]. The primary targets of fault attacks include cryptographic modules and critical combinational logic in the microprocessors. In recent works [5, 17, 45], it has been demonstrated that with deliberate fault attacks, the cryptographic keys can be recovered and the execution privilege of the microprocessor can be altered as well. Therefore, it becomes important to take the fault attack into consideration in the design stage.

In this chapter, both design and evaluation techniques are proposed to enhance the resilience against fault attacks. Specifically, a novel security primitive, PPUF is proposed in Sect. 4.2, which can work as a secure alternative of existing cryptographic modules and has demonstrated much stronger resilience against fault attacks. Meanwhile, in Sect. 4.3, an evaluation framework is proposed for the general microprocessors, which is able to evaluate the system vulnerability against fault attacks and identify the security-critical system components.

## 4.2 Practical PPUF Design

A PUF is a pseudo-random function that exploits the randomness inherent in the scaled CMOS technologies to generate unique output response given certain input challenge [13, 14, 21, 30]. A PPUF is a PUF that is created so that its simulation model is publicly available but large discrepancies exist between the execution



M. Li, D. Z. Pan, *A Synergistic Framework for Hardware IP Privacy and Integrity Protection*, https://doi.org/10.1007/978-3-030-41247-0_4

delay and simulation time [4, 22, 35, 38]. A PPUF relies on the time gap between execution and simulation to derive its security, which is promising because no secret information needs to be kept, and the enrollment phase before using a PUF (during which large amount of responses need to be characterized and stored) can also be eliminated [38]. Therefore, PPUFs are able to support multiple public-key protocols and have potentially much more applications compared with traditional PUFs [4].

For a PPUF to be an effective security primitive, ESG acts as a fundamental property and needs to be justified in terms of theoretical soundness and physical practicality. Theoretical soundness requires the ESG to be bounded rigorously, especially considering the advanced parallel and approximate computing scheme. Physical practicality further requires that the ESG can be realized effectively considering the existing fabrication techniques and the generated PPUF output must be measurable.

Although a number of PPUF designs have been proposed in the literature over the years [4, 27, 36–38], most of them do not justify the proposed ESG in the two aspects above. The first PPUF is proposed in [4] and relies on exclusive-or networks to convert the delay variation into small voltage glitches. Because the amount of glitches ideally increases exponentially relative to circuit depth, the authors claim that keeping record of all the glitches requires exponential computation. Although the idea is innovative, the PPUF is hard to realize because the generated glitches usually have very small pulse width and are very likely to be attenuated during propagation to PPUF output due to the electrical property of the logic gates [23]. Therefore, actual time gap is much smaller compared with the ideal expectation. Another security primitive, termed as SIMulation Possible, but Laborious (SIMPL) system, leverages the time gap between the real optical interference and solving the differential equations underlying the optical system [38]. However, its security relies on the nonlinearity of optical medium, which is still an open problem. To overcome the problem on optical medium, electrical implementations of SIMPL system based on memory cells and cellular nonlinear networks are proposed in [39]. However, rigorous proof on ESG and demonstration of PPUF circuit structure are not shown. In [37], a nano-PPUF design based on memristors is proposed. The authors justify the ESG by the complexity of matrix multiplication operation used in SPICE simulation. However, the authors ignore that matrix multiplication can be effectively paralleled to reduce the simulation time significantly.

While previous designs are more conceptual, we introduce a practical PPUF in Sect. 4.2. The PPUF execution is equivalent to solving the max-flow problem [16] in a complete graph, which enables us to use the max-flow problem as the simulation model and derive the upper bound of simulation time. We also derive the lower bound of the execution delay, based on which asymptotic bound of ESG can be proved rigorously. To enable an efficient physical realization of our design, we propose a crossbar structure to map the graph topology to silicon. The PPUF basic building block is designed with MOS transistors working in saturation region and further enhanced by source degeneration (SD) technique [29] to mitigate the short channel effect of transistors and instantiate flow constraints on chip. Design optimizations are proposed to further improve the practicality of the PPUF design. ESG is examined in experimental results by verifying the difference between asymptotic scaling of execution delay and simulation time. We summarize our contributions as follows:

- A new PPUF design is proposed with rigorous ESG achieved by solving max-flow problem in a complete graph on chip.
- A crossbar structure is proposed and enhanced with SD technique to map the graph topology and flow constraints on chip.
- Design optimizations are carried out to suppress the undesired factors and enable an efficient physical realization.

The rest of Sect. 4.2 is as follows. Section 4.2.1 describes preliminaries on max-flow problem and discusses the algorithms that aim to solve it. Section 4.2.2 introduces our PPUF topology and basic building blocks, which maps the max-flow problem on chip. ESG is also analyzed in Sect. 4.2.2. Section 4.2.3 describes the physical realization of the PPUF and also discusses the PPUF CRPs. We evaluate the performance of the PPUF in Sect. 4.2.4 and summarize in Sect. 4.2.5.

### *4.2.1 Preliminaries*

#### PPUF-Based Protocol

Let $S$ be a multiple-input-multiple-output physical system mapping challenges $c_i \in \mathscr{C}$ to responses $r_i \in \mathscr{R}$, with $\mathscr{C}$ and $\mathscr{R}$ denoting the finite set of all possible challenges. $S$ is called a PPUF if the following conditions are satisfied [38]:

- After presented with a challenge $c_i$, the corresponding response $r_i$ can be generated by $S$ within a pre-defined time threshold $t_{max}$.
- There exists a publicly available simulation algorithm $\mathsf{Sim}$ such that $\forall c_i \in \mathscr{C}$, the algorithm $\mathsf{Sim}$ outputs $r_i$ in feasible time.
- Any adversary Eve bounded by practically feasible computations will succeed in the following security experiments with a probability of at most $\epsilon$:
  - Eve is given the simulation algorithm $\mathsf{Sim}$.
  - Eve is presented with a challenge $c_c$ that is uniformly and randomly chosen from $\mathscr{C}$, and is asked to output a value $v_c$.

  Here we define that Eve succeeded in the security experiments if

  - $v_c = r_c$, where $r_c$ is the correct response of $S$.
  - The time that Eve needed to output $v_c$ after $c_c$ was presented is less than $t_{max}$.

To satisfy the requirements above, PPUFs not only need to be immune to physical clone or machine-learning-based functional clone [40] like regular PUFs, but also need to have a provable timing gap between physical execution and simulation. The

**Protocol 1** Entity identification based on PPUF

1: Bob acquires Sim, $t_{max}$ and $\mathscr{C}$ associated with Alice.
2: Bob randomly picks $c_c \in \mathscr{C}$ and sends it to Alice.
3: Alice queries her PPUF $S$ to determine the response $v_c$ and sends back to Bob.
4: Bob receives $v_c$ and verifies Alice's response time $t$. If $t > t_{max}$, Bob denies Alice's identity; while if $t < t_{max}$, Bob checks the correctness of $v_c$ by Sim. If $v_c = r_c$, Bob believes Alice's identity, otherwise, he denies Alice's identity.

ESG needs to be justified rigorously considering the most advanced computation algorithms and resources as well as the practical of physical realization.

Because Sim is available to the public, the enrollment phase in protocols based on traditional PUFs can be eliminated, which enables PPUF to support more applications and protocols compared with traditional PUFs. We take the most widely used protocol of entity identification as an example. Assume that Alice holds the PPUF $S$ and has released Sim, $t_{max}$ and a description of $\mathscr{C}$ to the public. Not, Alice can prove her identify to an arbitrary second party Bob following the protocol shown in Protocol 1.

As shown in Protocol 1, due to the publicly available simulation model and the ESG, there is no need for Bob to characterize the challenge-response relation of Alice's PPUF and save the CRPs before the identification procedure. Therefore, ESG is the most fundamental property of PPUF. In Sect. 4.2.2, we will describe our PPUF design and prove the asymptotic bound for the ESG rigorously.

#### Max-Flow Problem in Directed Graph

Let $G = (V, E)$ represent a directed graph with $|V| = n$ vertices and $|E| = m$ directed edges. If $\forall v, u \in V, \exists (v, u) \in E$, $G$ is a complete graph with $m = n(n-1)$. In the directed graph $G$, we distinguish a set of source vertices $S \subset V$ and sink vertices $T \subset V$ and assign a non-negative capacity $c(v, u)$ to each edge $(v, u) \in E$. An instance of the max-flow problem consists of the directed graph $G$ and the set of capacities.

Given an instance of a max-flow problem, a function, $f : E \rightarrow R^+$, is called a flow function if it assigns a non-negative value to each edge that satisfies the conservation constraint

$$\sum_{(v,u)\in E} f(v, u) = \sum_{(u,v)\in E} f(u, v) \qquad \forall v \in V \setminus (S \cup T),$$

and the capacity constraints

$$0 \leq f(v, u) \leq c(v, u) \qquad \forall (v, u) \in E.$$

The value of a flow $f$ is defined as the net flow from a source node. The max-flow problem is to find a maximum-value flow function on a given instance of a flow problem.

Max-flow problem has been shown to be computationally demanding and difficult to parallel or approximate [16]. Traditional methods include augmenting-path algorithm[10], push-relabel algorithm[15], blocking flow method [10], and so on. All these methods have at least $O(n^3)$ complexity for complete graph. Recent efforts on solving max-flow problem efficiently can be classified into two categories, including parallel methods and approximate methods. The best known parallel method shown in [42] leverages blocking flow algorithm and achieves a parallel runtime of $O(n^3\log(n)/p)$, where $p \leq n$ is the number of processors. Therefore, the best achievable complexity of the parallel algorithms is lower bounded by $O(n^2\log(n))$. The best known approximate algorithm targeting at max-flow problem is proposed in [20]. To get an $\epsilon$-approximate solution, the complexity of the proposed algorithm is $O(m^{1+o(1)}\epsilon^{-2})$, which becomes $O(n^{2+o(1)}\epsilon^{-2})$ for complete graph. Therefore, considering the parallel and approximate algorithms, the complexity of solving max-flow problem is still lower bounded by $O(n^2)$.

While solving the max-flow problem is computationally intensive, it is much easier to check the optimality of a flow $f$. Define the residual capacity $r_f(v,u)$ of an edge $(v,u)$ to be $c(v,u) - f(v,u)$. Then, the residual graph $G_f = (V_f, E_f)$ can be built for a flow $f$, which is a directed graph with $V_f = V$ and $E_f = \{(u,v) \in E : r_f(u,v) > 0\}$. The flow function $f$ is optimal if and only if $\forall t \in T, s \in S, t$ is not reachable from $s$ in the residual graph. Therefore, to check optimality of a flow solution, we just need to create the residual graph and do a breadth-first search from source to sink, which is highly parallelizable and can be finished with $O(n^2/p)$ complexity for a complete graph [47].

*Example 4.1* Consider the graph $G = (V, E)$ as shown in Fig. 4.1a. We have $V = \{s, v_1, v_2, v_3, t\}$, $S = \{s\}$, $T = \{t\}$ and $E = \{(s,v_1), (s,v_2), (v_2,v_1), (v_3,v_2), (v_1,v_3), (v_1,t), (v_3,t)\}$. The capacity and flow of each edge is annotated in the graph. In the graph, consider $(v_1,t) \in E$, the flow on the edge is 10, which does not exceed the capacity. For $v_1 \in V$, the flow into $v_1$ is 14 and equals to the flow out

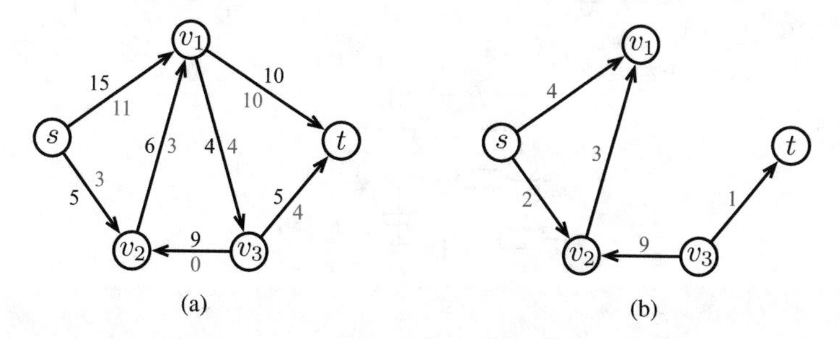


**Fig. 4.1** Example of max-flow problem in directed graph: (**a**) original graph with capacity (black) and flow (red) for each edge; (**b**) residual graph with residual capacity (blue)

of $v_1$. The residual graph as well as the residual capacity is shown in Fig. 4.1b. In the residual graph, no path exists from source $s$ to sink $t$, which indicates the current flow function achieves the maximum value, which equals to 14.

### Signal Delay in General Circuit

To prove the ESG rigorously, execution delay of the PPUF needs to be bounded. Although Elmore delay can provide a sufficient estimation for the execution time [25], it is still not easy to evaluate since our PPUF structure forms a complete graph. Different from traditional tree structure, for which Elmore delay can be evaluated easily, the driving and loading networks of a circuit node in our PPUF are not explicit. In fact, all circuit nodes are coupled together in the graph, so that every node is driving and loading other nodes at the same time. To evaluate the signal delay for general graph, authors in [25] propose a method to determine the delay values of all the nodes collectively.

The main intuition of [25] is illustrated in Fig. 4.2. Consider a vertex $v \in V$ with $l$ neighbors $u_1, \ldots, u_l$. Let $R(u_i, v)$ denote the resistance of the edge between $u_i$ and $v$ and $C(v)$ denote the capacitance of $v$ as shown in Fig. 4.2a. If we decompose $C(v)$ into $l$ parts and redistribute each part to the edges that connects $u_i$ and $v$, denoted as $C(u_i, v)$, as in Fig. 4.2b, then, the current on each $(u_i, v)$ as well as the delay for each $u_i$ and $v$ are not changed. If we further require the voltages of $C(u_i, v)$, $1 \leq i \leq l$ to be the same, then, there is no current between the connections of different $C(u_i, v)$, which indicates in Fig. 4.2c, the current on each $(u_i, v)$ as well as the delay for each $u_i$ and $v$ remain the same as Fig. 4.2a. The method proposed

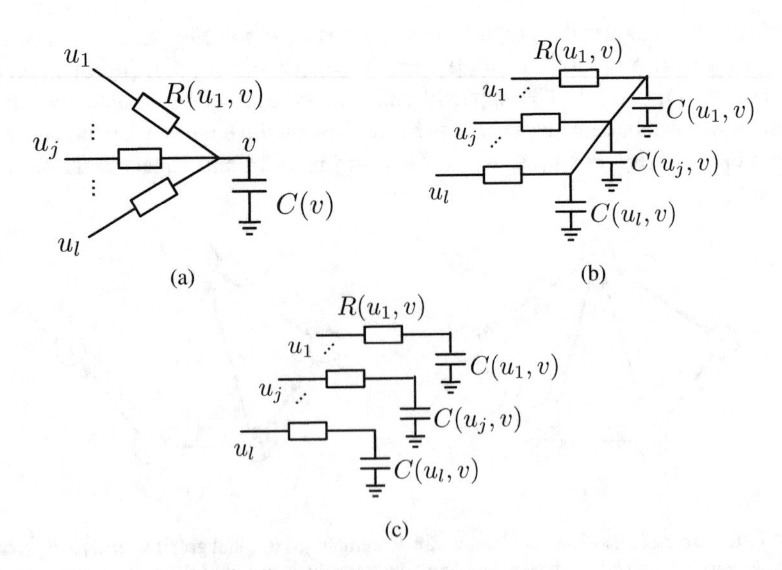


**Fig. 4.2** Capacitance decomposition and redistribution

in [25] calculates the Elmore delay of general circuit by selecting circuit nodes to decompose and redistribute the capacitance iteratively.

To enable the capacitance decomposition and redistribution, the following relations hold [25]

$$T(v) = T(u_i) + R(u_i, v)C(u_i, v) \qquad \forall 1 \le i \le l$$

$$\sum_{1 \le i \le l} C(u_i, v) = C(v)$$

Here, $T(u_i)$ denotes the delay from source node to $u_i$. To be noticed here, $C(u_i, v)$ is just a mathematical representation, which can be either positive or negative depending on the relation between $T(u_i)$ and $T(v)$. As proved in [25], by solving the linear equations, delay for each node can be accurately calculated.

### 4.2.2 PPUF Topology and ESG Analysis

In this section, we introduce our PPUF design and rigorously prove the ESG. Our main intuition is to build a PPUF circuit whose execution is equivalent to solving the max-flow problem but requires asymptotically less time compared with best known algorithms. The main difficulty comes from mapping the constraints and objective functions on chip. As we will show, our PPUF topology together with the basic building block guarantees the equivalence,and thus, enables a rigorous ESG.

#### PPUF Topology and Basic Building Block

The proposed PPUF topology is shown in Fig. 4.3. The PPUF consists of a pair of nominally identical networks that are different only because of process variation. The circuit nodes correspond to the vertices in the graph, while each building block as shown in Fig. 4.4d instantiates one directed edge. Inputs to the PPUF are used to select the source nodes and sink nodes, and control the current capacity of each edge. The selected source and sink nodes are connected to $V(s)$ and ground, respectively. Output is generated by comparing the current flowing into the source node.

To explain our design methodology for the basic building block, we list our requirements below and describe the proposed circuit block step by step to ensure all the requirements are satisfied.

**Requirement 1** *The maximum current of the basic building blocks must be controllable.*

This is because the basic blocks are used to instantiate edges in the max-flow problem, for which capacity constraints are enforced. To satisfy the requirement, we use MOS transistors working in saturation region as in Fig. 4.4a, and control

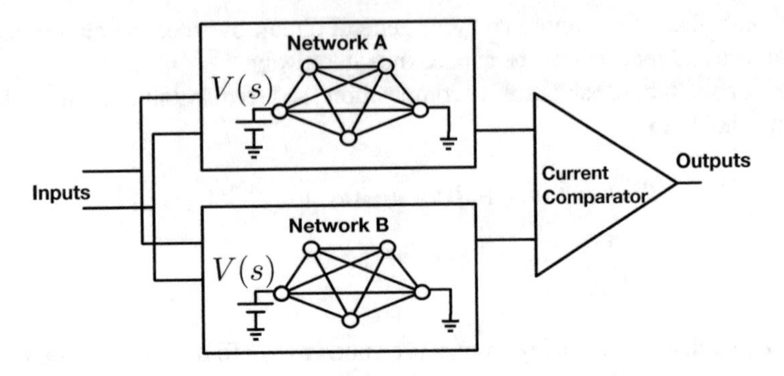


**Fig. 4.3** Topology of the proposed PPUF design

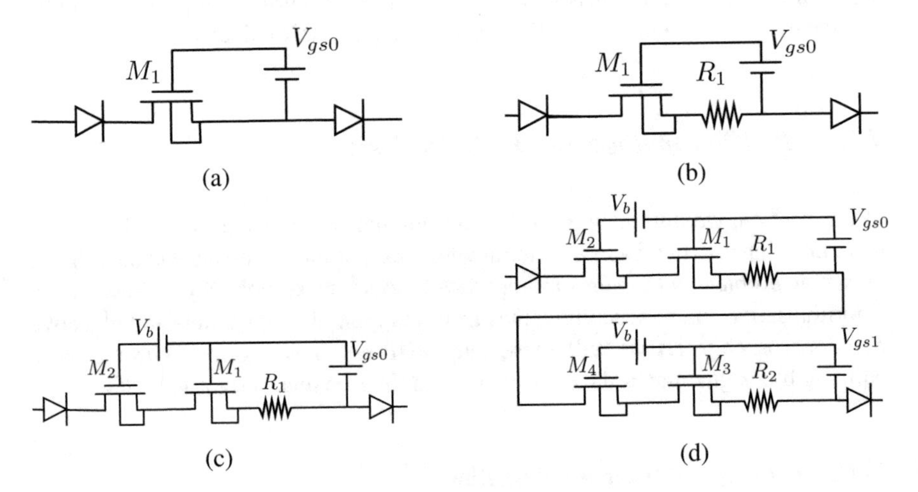


**Fig. 4.4** Evolution of basic building block design to satisfy all the requirements

the maximum current by setting the gate-to-source bias $V_{gs}$. The diodes are used to ensure the direction of the current, corresponding to the direction of each edge in the graph.

We show the $I$–$V$ characteristic of the basic block with the change of drain-to-source bias $V_{ds}$ in Fig. 4.5a (blue line). As we can see, given fixed $V_{gs}$, the saturation current still changes significantly, which is due to the short channel effects (SCEs) for MOS transistors. The change of saturation current due to SCEs determines the discrepancy between the PPUF execution and the corresponding simulation model, i.e. max-flow problem, and thus, needs to be suppressed. To mitigate the impact of SCEs and reduce the change of saturation current, we adopt the SD technique that is widely used in analog circuit design. SD technique can help stabilize the current and mitigate the impact of SCEs by introducing negative feedback with resistors or MOS transistors. As in Fig. 4.4b, $R_1$ acts as the degeneration resistor for $M_1$. After $M_1$ enters the saturation region, the change of current caused by the increase of

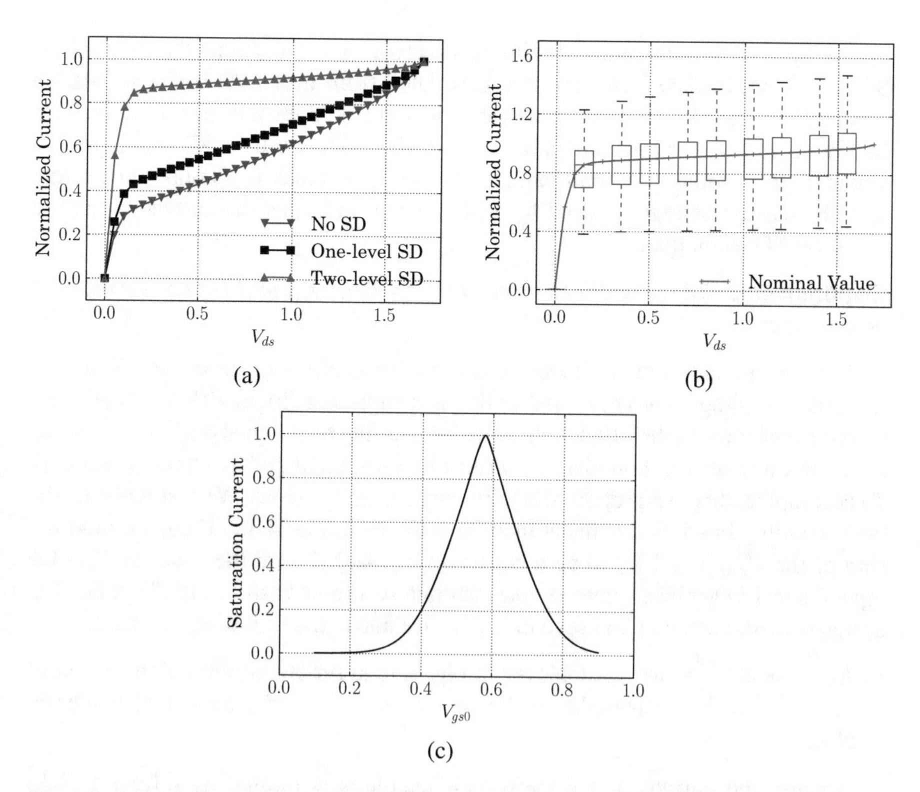


**Fig. 4.5** $I$–$V$ relation of proposed circuit unit: (**a**) comparison on saturation current change for different basic block designs; (**b**) impact of process variation on basic block design; (**c**) relation between saturation current and control voltage

$V_{ds}$ can be compensated by the increased voltage drop on $R_1$. We show its current characteristics in Fig. 4.5a (black line). As we can see, the change of saturation current is reduced due to the SD technique.

The degeneration technique can be nested to further suppress the change of saturation current. As in Fig. 4.4c, two levels of source degeneration are nested: $R_1$ works as the degeneration circuit for $M_1$, while $M_1$ and $R_1$ together work as the degeneration circuit for $M_2$. The additional voltage source ($V_b$) is used to ensure both $M_1$ and $M_2$ are working in saturation region. As shown in Fig. 4.5a (red line), with nested two-level SD technique, the change of saturation current due to SCEs is mitigated significantly. In fact, we can gain better control over saturation current with more levels of SD technique, which will also lead to larger design overhead. To decide the sufficiency of the SD technique, we have the following requirement.

**Requirement 2** *The impact of process variation on the saturation current needs to be much larger compared to the discrepancy induced by SCEs.*

During PPUF execution, output response is generated by comparing two nominally identical current that are different only because of process variation. Therefore, by the second requirement, we aim to ensure that the discrepancy induced by SCEs will not lead to false response calculated from max-flow based simulation compared to real PPUF execution. As shown in Fig. 4.5b, experimental results from SPICE based Monte Carlo simulation indicate that with two-level SD technique, the amplitude of saturation current variation of the basic block is around 100× larger than the current change induced by SCEs, which indicates the sufficiency of the two-level SD technique.

**Requirement 3** *The boundary between PPUF 0-response and 1-response needs to be nonlinear.*

The requirement aims to ensure good resilience of the proposed PPUF design to model-building attacks. Model-building attacks aim to model the challenge-response behavior of the PPUF with machine learning techniques [40]. To ensure the resilience, a nonlinear boundary between PPUF 0-output and 1-output is required. To accomplish this, we propose the following design heuristic. We first replicate the basic building block and connect them in serial as in Fig. 4.4d. Then, we limit the sum of the $V_{gs0}$ and $V_{gs1}$ to be a constant ($V_c$), and choose the value of $V_{gs0}$ for input 0 and 1 to achieve same nominal saturation current as shown in Fig. 4.5c. The design heuristic can help improve the resilience due to the following two reasons:

- After connecting two basic blocks in Fig. 4.4c in serial, the saturation current of the basic block equals to the smaller saturation current of the previous two basic blocks.

Because the current of the basic building block is limited by different MOS transistors for input 0 and 1, given the current information for input 0, the current information for input 1 remains unknown and vice versa. Meanwhile, because the current of all the edges connected to the node sum up to 0 for each internal node, the current flowing through one edge is not only determined by the voltage of the edge, but also impacted by all the other edges connected to the node. Therefore, all the inputs are closely correlated to achieve a nonlinear boundary between 0-output and 1-output. The requirement is also verified in the experimental results with both parametric and non-parametric model-building techniques.

Besides satisfying all the requirements above, another intriguing property of the building block is its incremental passivity [28]. A memoryless component is incrementally passive if its current increases monotonically as the increase of voltage. The proposed building block satisfies this condition. As we will show, the incremental passivity of basic block helps ensure that the steady state current of the PPUF circuit is optimal solution to the corresponding max-flow problem.

#### Lower Bound of PPUF Simulation

In this section, we will prove the equivalence between execution of PPUF and solving max-flow problem in a complete graph. The equivalence will enable us to use the max-flow problem as the simulation model and derive the lower bound of the simulation time.

First consider the basic block that points to circuit node $u$ from node $v$. Let $I(u, v)$ denotes the current of the basic block. If we define the current to be positive and if it flows into node $v$, then the diodes on the two sides of the basic block guarantee $I(u, v)$ is always non-negative. Meanwhile, given fixed control voltage $V_{gs0}$, the current of the basic block is limited by the saturation current $I_{sat}(u, v)$ as we have shown above. Therefore, we have

$$0 \le I(u, v) \le I_{sat}(u, v), \qquad \forall (u, v) \in E.$$

Then, consider circuit node $v \in V \setminus (S \cup T)$. Based on Kirchhoff's current law, the current flowing into $v$ should always be the same as the current flowing out of $v$, i.e.

$$\sum_{(v,u)\in E} I(v, u) = \sum_{(u,v)\in E} I(u, v)$$

Now consider the circuit node $s$ that connects to the supply voltage in PPUF as shown in Fig. 4.6. Based on Kirchhoff's current laws, we have

$$I(s) = \sum_{(s,v)\in E} I(s, v),$$

where $I(s)$ represents the current flowing into $s$ from the supply voltage. Intuitively, with the increase of $V(s)$, $I(s)$ will also increase. Due to the current constraints as derived above, $I(s)$ eventually gets saturated and such saturated $I(s)$ is the maximum possible flow into the PPUF.

More formally, we rely on the concept of incremental passivity [28] to prove the intuition. A memoryless circuit unit is incrementally passive if its current increases monotonically with the voltage on the circuit unit. Following the definition, our basic block is incrementally passive. Since the whole PPUF circuit only consists of such basic building blocks, it is also incrementally passive [28]. The property helps us to guarantee the following two properties of the PPUF circuit:

- With the increase of $V(s)$, the current flowing into the PPUF circuit increases monotonically.

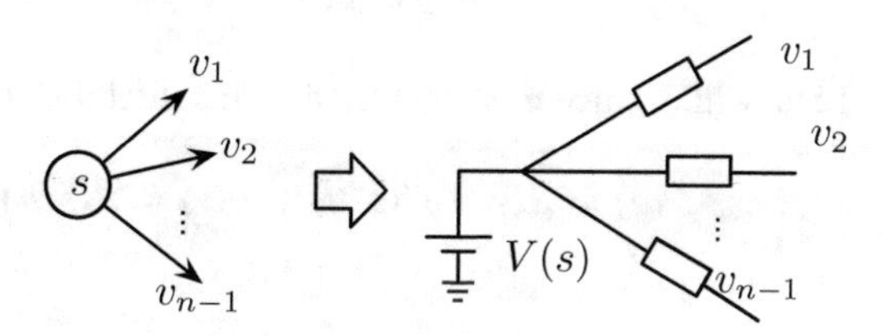


**Fig. 4.6** PPUF circuit mapping to objective function

- Regardless of the initial condition, the current of the PPUF will converge to a unique solution given fixed input voltage.

Therefore, by increasing $V(s)$, $I(s)$ will be maximized under the current constraints.

Based on the analysis above, our PPUF design is equivalent to the following optimization problem

$$\begin{aligned} &\textbf{max } I(s) \\ &\textbf{s.t. } 0 \leq I(v,u) \leq I_{sat}(v,u), && \forall (v,u) \in E; \\ &\sum_{(v,u)\in E} I(v,u) = \sum_{(u,v)\in E} I(u,v), && \forall v \in V \setminus (S \cup T). \end{aligned}$$

Therefore, we can conclude that the PPUF execution is equivalent to solving a max-flow problem in a directed graph. Specifically, since each node in the PPUF is designed to be connected with all the other nodes, the direct graph is complete. The equivalence enables us to use the max-flow problem as the simulation model. More importantly, because the max-flow problem is hard to parallel and approximate, we are able to rigorously derive the lower bound for the simulation time: with the best known algorithm, the simulation time scales at least $O(n^2)$ as the increase of PPUF node number.

### Upper Bound of PPUF Execution

To derive the upper bound of PPUF execution, consider the node with largest delay in the PPUF network, denoted as $u$. Then, from the linear equations described in Sect. 4.2.1, by capacitance decomposition and redistribution, the delay of $u$, denoted as $T(u)$, satisfies

$$\begin{aligned} T(u) &= T(v) + R(v,u)C(v,u) \qquad \forall v \in V \setminus \{u\} \\ C(u) &= \sum_{(v,u)\in E} C(v,u) \end{aligned}$$

Since we assume $u$ has the largest delay, we have $T(u) \geq T(v), \forall v \in V - \{u\}$. Therefore, the redistributed capacitance satisfies

$$0 \leq C(v,u) \leq C(u) \qquad \forall v \in V \setminus \{u\}$$

Meanwhile, since $u$ is connected with $s$ directly, we also have

$$T(u) = R(s,u)C(s,u) + T(s) = R(s,u)C(s,u) \leq R(s,u)C(u)$$

Here $R(s, u)$ is the resistance of the edge connecting $s$ and $u$, which remains unchanged as the increase of node number. $C(u)$ is the capacitance of node $u$, which increases linearly because the number of edges incident on $u$ increases linearly. Therefore, the delay for the PPUF scales at most $O(n)$ with respect to circuit node number.

### ESG Amplification

The analysis in Sects. 4.2.2 and 4.2.2 rigorously proves that ESG exists considering the parallel and approximate computing scheme. To further increase the ESG, we propose to leverage the methods as proposed in [38]. According to [38], the verifier can present a challenge $C_1$ to the PPUF holder or attacker but force him to evaluate a sequence of challenge-response pairs $(C_1, R_1), \ldots, (C_k, R_k)$, in which later challenges $C_i$ is determined by earlier responses $R_{i-1}$, where $2 \leq i \leq k$. In this way, $C_1, R_k$ is regarded as the overall challenge-response pair.

The method is termed as feedback loop technique. Because $C_i$ depends on $R_{i-1}$, the PPUF holder or the attacker is forced to evaluate the $k$ challenge-response pairs in serial. In this way, the lower bound of the simulation time becomes $O(kn^2)$, while the upper bound of the PPUF execution delay becomes $O(kn)$ and thus, ESG is amplified by $k$ times.

### Speeding Up Verifier's Task

According to the PPUF-based authentication protocol [38], verifiers have the same simulation model as the attackers. To determine the correctness of the response, direct simulation of the proposed PPUF by solving the max-flow problem can impose same computation burden for the verifiers compared with attackers, which is undesired. However, since the task for verifiers is to verify the correctness of the response, they may not need to carry out the actual simulation. Our main intuition is to leverage the computational asymmetry between solving the max-flow problem and verifying the correctness of a solution. To accomplish this, each time after receiving the response, the verifier can further ask for the information on the voltage of each circuit node in the PPUF, which can be measured directly. Based on the voltage of circuit nodes, the current of each basic building block can be calculated by the verifier. Equivalently, the verifier can get the flow on each edge in the max-flow problem, based on which the residual graph can be built for the max-flow problem. Given the residual graph, the verifier just need to use a breath-first search starting to from source nodes to determine whether the sink nodes can be reached. If the sink node is not reachable, then, the flow is the optimal solution and the response from the PPUF holder or attacker is considered correct. The verification process can be finished in one traversal of the whole graph and can be highly parallelizable as described in Sect. 4.2.1, which significantly reduces the computation burden for the verifier.

### 4.2.3 PPUF Physical Realization

Although the ESG is proved rigorously in Sect. 4.2.2, realizing a complete graph and the basic building block on chip is non-trivial. In this section, we describe our strategies towards a practical and efficient on-chip realization of the proposed PPUF design.

#### Complete Crossbar Structure

The completeness of the PPUF structure requires each circuit node to be connected with all the other nodes. To realize the complete connection, we propose a novel crossbar structure. As shown in Fig. 4.7b, the number of horizontal and vertical bars are the same as the node number. The $i$th horizontal bar and $i$th vertical bar are connected directly through a wire, which represent one node in the PPUF circuit. Then, at the intersection of $i$th vertical bar and $j$th horizontal bar ($i \neq j$), there is a basic building block. The direction of the building blocks is always pointing from the vertical bars to the horizontal bars. In this way, each bar is connected with all the other bars through the basic building blocks, which realizes the complete connection of vertices in the graph. We use the following example to illustrate the crossbar structure in detail.

*Example 4.2* Consider a PPUF circuit, whose execution is equivalent to solving the max-flow problem in the graph as in Fig. 4.7a. The corresponding crossbar structure is shown in Fig. 4.7b. As in Fig. 4.7b, the first horizontal and vertical bar are connected through a wire, which instantiates the source node $s$ in Fig. 4.7a and thus, is connected to $V(s)$. The second horizontal bar and the fourth vertical bar, which are parts of node $v_1$ and $t$, respectively, are connected through a basic block. This instantiates the edge $(v_1, t)$ in Fig. 4.7a. The direction of the basic block is always pointing to the vertical bar from the horizontal bar. By examining the one-

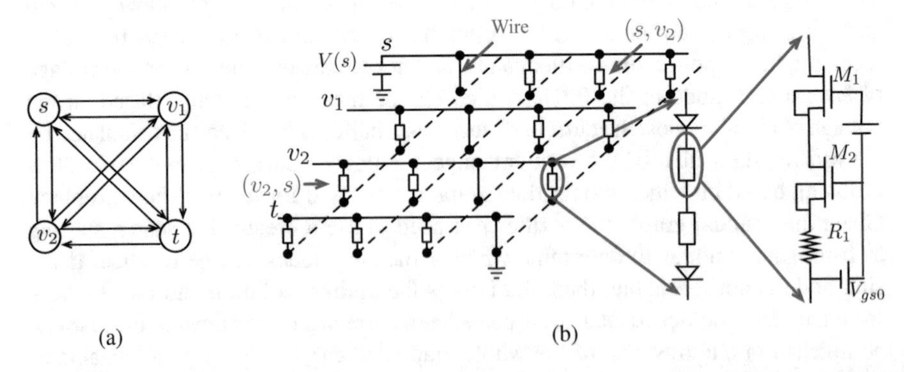


**Fig. 4.7** Crossbar structure to map the complete graph on chip: (**a**) example graph; (**b**) crossbar structure that instantiates graph in (**a**)

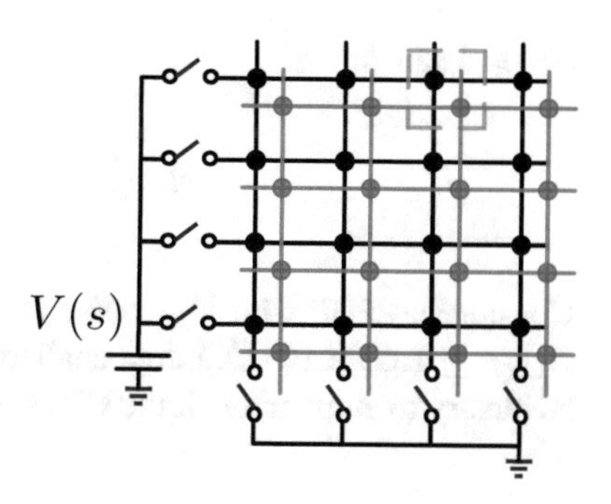


**Fig. 4.8** Crossbar placement and type-A control inputs: crossbar A (black) and crossbar B (red) are placed interleavingly with basic blocks in the same positions placed side by side (green box). Type-A inputs determine the connection of each node with ground and supply voltage

to-one correspondence of the wires and basic blocks, we can verify that the crossbar structure correctly implements the PPUF circuit in Fig. 4.7a.

### Interleaving Crossbar Placement

To ensure enough ESG, the circuit size of the PPUF can become large. In this context, systematic variation becomes non-negligible, which may lead to biased and detectable challenge-response behavior for the PPUF. To mitigate the impact of systematic variation, we propose an interleaving placement strategy. As shown in Fig. 4.8, the two nominally designed networks are placed interleavingly. The transistors in the same positions from the two different networks are placed side by side. In this way, transistors in the same positions can be assumed to have the same systematic variation. Due to the differential structure of the proposed PPUF, the impact of systematic variation can be suppressed significantly.

### Input Challenge Pre-pruning

To satisfy the requirement on output flip probability, we propose to prune the input challenge space before the deployment of the PPUF circuit. Our strategy is to select a subset from the whole challenge space such that the minimum Hamming Distance for different challenges is larger than a pre-defined threshold $d$. This means for any two challenge inputs, at least $d$ type-B challenge bits are different, and thus, the maximum current for at least $d \times \frac{m}{l^2}$ basic blocks are different, which greatly increases the probability for a bit flip. By controlling $d$, as we will show in the experimental results, the output flip probability can approach 0.5.

Now we can analyze the number of challenge-response pairs for our PPUF design. For type-A input, to select one source and sink nodes, we can have totally $n(n-1)$ different choices. For type-B input, to decide the challenges that satisfy the requirement on the minimum distance $d$, it is equivalent to constructing binary codes of length $L^2$ and minimum HD $d$. As proved by [34], the size of the type-B challenge space is larger than $2^{L^2}/(\sum_{i=0}^{d-1}\binom{L^2}{d})$. Then the total number of CRPs

($N_{CRP}$) satisfies

$$N_{CRP} \geq n(n-1) \times \frac{2^{L^2}}{\sum_{i=0}^{d-1} \binom{L^2}{d}}$$

Consider a PPUF with $n = 400$ circuit nodes. Assume $L = 20$ and $d = 2L$, then $N_{CRP} \geq 1.68 \times 10^{+71}$. Large challenge-response space makes it impossible for an adversary to enumerate all the CRPs exhaustively.

### 4.2.4 Experimental Results

In this section, we examine the security properties of the proposed PPUF design. The experiments fall into the following categories: accuracy of the simulation model, asymptotic scaling of ESG, PPUF output measurability and power consumption, statistical evaluation of PPUF metrics, and model-building attack resilience.

The current output and execution delay of the PPUF circuit is acquired using SPICE simulation with 32 nm predictive technology model [1]. We assume the threshold voltage variation follows normal distribution with a standard deviation of 35 mV, a value consistent with ITRS [19]. Concerning the voltage settings, $V(s)$ is set to be 1.2 V since we consider ideal diode. We also set $V_b = 0.1$ V, $V_c = 1.2$ V. If input is 1, $V_{gs0}$, as shown in Fig. 4.4d, is set to be 0.5 V, while if input is 0, $V_{gs0}$ is set to be 0.67 V. The simulation model is implemented in C++. Because the best known sequential and parallel algorithms are more conceptual with no packages available, we instead choose the most widely used push-relabel and augmenting-path algorithms from boost library [6]. Although it is possible to reduce the simulation time by running on better machine or using more efficient algorithm, we argue that the lower bound of the simulation time still exists and justified ESG can be guaranteed as we have proved. Meanwhile, because the statistical evaluation for PPUFs with large number of nodes is too time-consuming, we demonstrate most of the tests on relatively small PPUFs and use extrapolation to estimate the performance for large PPUF. We run all the experiments on an Intel Xeon 2.93 GHz workstation with 74G memory.

#### Verification of Simulation Model and ESG

We first demonstrate the accuracy of using max-flow problem as the simulation model. We compare the results from execution and simulation for PPUFs with different number of nodes and define the inaccuracy as $|I_{max,exe} - I_{max,sim}|/I_{max,exe}$. For each PPUF, we run 100 simulations and show the inaccuracy in Fig. 4.9. As we can see, the average inaccuracy is less than 1%. Compared with the inaccuracy, the average variation of the maximum current flow is around 9.27% for a 100-

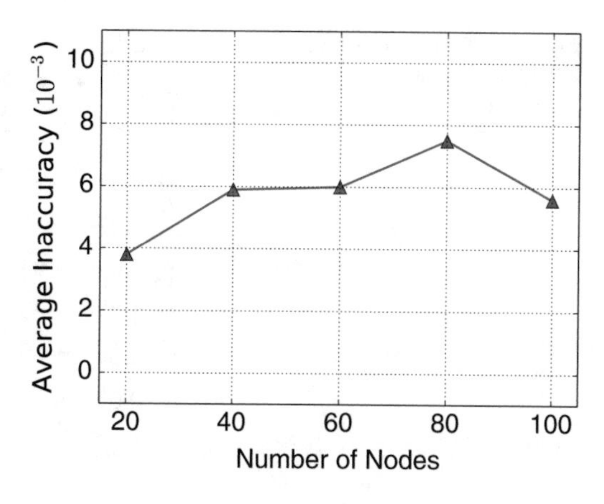


**Fig. 4.9** Inaccuracy of simulation model compared with PPUF execution

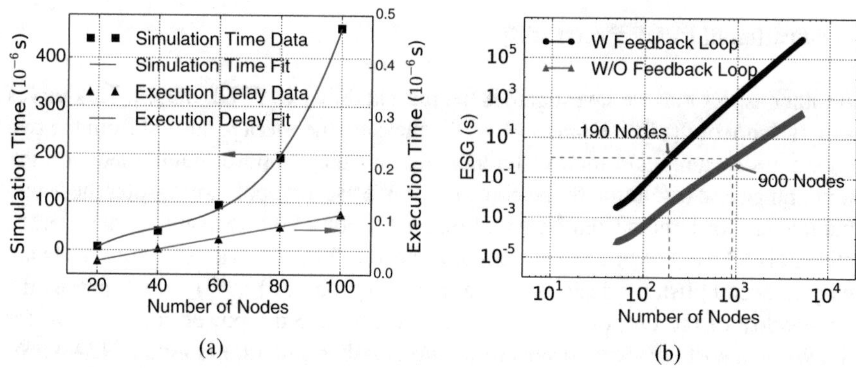


**Fig. 4.10** Comparison between execution and simulation time: (**a**) scaling of execution and simulation time and polynomial fitting; (**b**) scaling of ESG with/without feedback loop technique

node PPUF. The comparison ensures that we can get accurate response from the simulation model.

Next, we demonstrate the ESG by comparing the PPUF execution and simulation time. The scaling of execution delay and simulation time is shown in Fig. 4.10a. Then, ESG can be calculated as the difference between the execution delay and simulation time. We show the ESG with/without feedback loop technique in Fig. 4.10b. For feedback loop technique, we set the loop number to be the same as the node number in PPUF, i.e. $k = n$. As we can see, to achieve 1s ESG, which is shown to be a reasonable requirement in [35], 900 nodes are needed for our PPUF design, while with feedback loop technique, the required number of nodes reduces to 190.

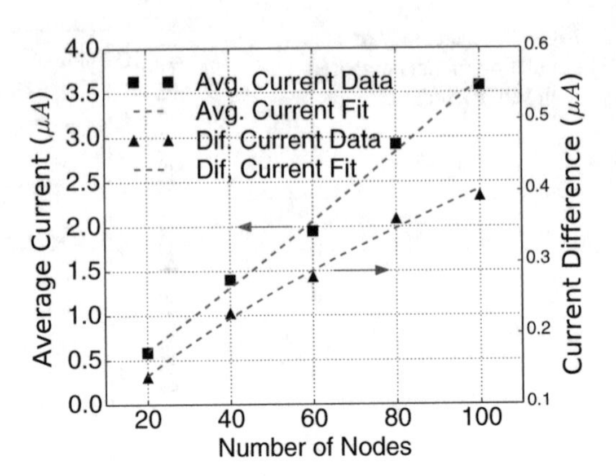


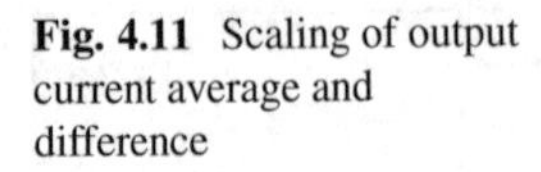
**Fig. 4.11** Scaling of output current average and difference

### Verification of PPUF Practicality

Another aspect that we investigate is the measurability of PPUF output. This serves as a measure of the PPUF practicality. We measure the average current from the two crossbar structures and their difference because they impose requirements on the input range and resolution of the comparator. We use extrapolation to infer these two parameters for large design based on Fig. 4.11. For a 900-node PPUF, the average current is 33.6 μA, while the current difference is 2.89 μA. These requirements are easy to accomplish by designs shown in existing papers [7, 43], which proves the practicality of the proposed design. We also estimate the power consumption for the 900-node PPUF. The power of the two crossbar structure is around 134.4 μW. As for the current comparator, we use the data from [43], which is 153 μW. Based on Fig. 4.10a, the execution delay for a 900-node PPUF is estimated to be 1.0 μs. Therefore, for one evaluation, the total power consumption is around 287.4 pJ.

### PPUF Statistical Evaluation

We further examine the PPUF performance over several commonly used metrics that quantify the quality of the PPUF design: inter-class HD, intra-class HD, randomness, and uniformity [26]. In our experiments, intra-class HD accounts for supply voltage variation of 10% and temperature variation ranging from −20 °C to 80 °C. We evaluate these metrics for a 40-node and a 100-node PPUF. As we can see in Table 4.1, the average performance of both PPUFs is close to ideal value.

We also evaluate the relation between output flip probability and minimum HD ($d$) of PPUF challenge: changing $d$ inputs, we check the probability for the output bit to flip. Here we run experiments on 100 40-node PPUF circuits with gird size $l = 8$. For each PPUF and each minimum HD $d$, we random sample 1000 input vectors. The change of output flip probability relative to $d$ is shown in Fig. 4.12. As we can see, when $d = 16$, the average output flip probability approaches to 0.5.

**Table 4.1** Statistical evaluation on 40-node and 100-node PPUF

| Metrics | Ideal | 40-Node PPUF | | 100-Node PPUF | |
|---|---|---|---|---|---|
| | | Mean | Stdv | Mean | Stdv |
| Inter-class HD | 0.5 | 0.5009 | 0.1371 | 0.4977 | 0.1075 |
| Intra-class HD | 0 | 0.0673 | 0.1104 | 0.0853 | 0.1321 |
| Uniformity | 0.5 | 0.4946 | 0.208 | 0.4672 | 0.158 |
| Randomness | 0.5 | 0.4946 | 0.0277 | 0.4672 | 0.0361 |

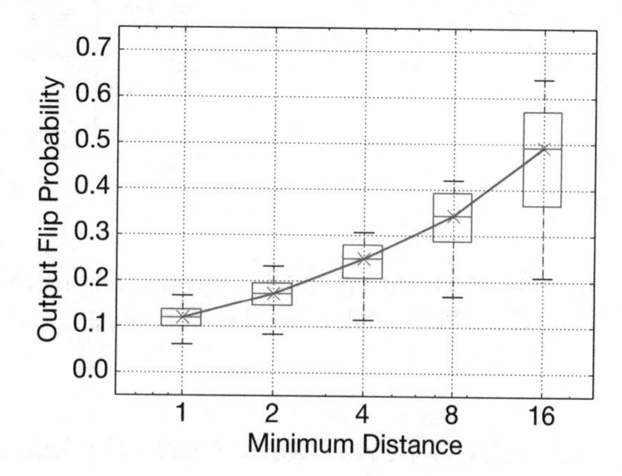


**Fig. 4.12** Output bit flip probability with respect to minimum distance of input challenges

#### Model-Building Attack Resilience

To evaluate the model-building attack resilience, we leverage both parametric and non-parametric machine learning algorithms, including support vector machines (SVMs) [44] and $K$ nearest neighbor (KNN) [9]. We employ a nonlinear radial basis function (RBF) kernel for SVM algorithm, while for KNN algorithm, we run a series of empirical KNN tests with $K = 1, 3, \ldots, 21$. The final prediction inaccuracy is the minimum of SVM and KNN tests. The prediction error for 40-node and 100-node PPUFs is shown in Fig. 4.13. Compared to the arbiter PUF with the same input length, our PPUF achieves more than an order of magnitude higher prediction error than arbiter PUF, which indicates much better model-building attack resilience.

### *4.2.5 Summary*

In Sect. 4.2, we propose a PPUF with practical ESG in terms of theoretical soundness and physical practicality. The execution of PPUF is proved to be equivalent to calculating max-flow in a complete graph, which enables us to use the max-flow problem as the simulation model to rigorously bound the simulation time. The execution time is also bounded for the proposed design. Therefore, rigorous ESG can be shown based on the difference on the asymptotic scaling. To enable an efficient realization of PPUF, we propose a crossbar structure and adopt the SD

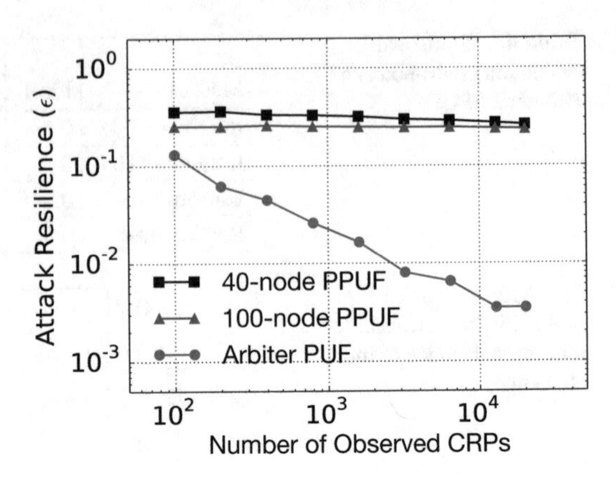


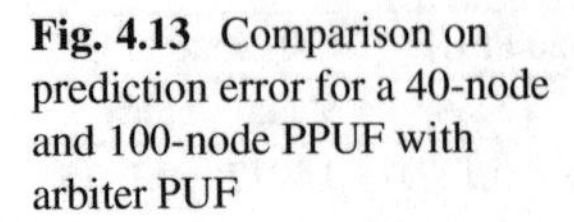

**Fig. 4.13** Comparison on prediction error for a 40-node and 100-node PPUF with arbiter PUF

technique to build the PPUF basic building blocks to map the complete graph on chip. Our PPUF exhibits good performance as shown in the experimental results.

## 4.3 Cross-Level Monte Carlo Evaluation Framework

Several methods have been proposed to evaluate the system vulnerability against fault attack in recent papers [33, 41, 50]. In [33], the authors propose a framework that is able to identify and mitigate fault attack vulnerability in the finite state machine (FSM). By extracting the state transition and identifying the don't-care states, the vulnerability of FSM against fault attacks can be analyzed. In [50], the authors analyze the system vulnerability against attacks that cause timing violation in the circuits. By considering the fault generation and propagation, accurate fault impact can be modeled for combinational logic. In [8, 11, 45], fault attack analysis is proposed for cryptographic modules including RC4, AES, DES, and so on, based on the mathematical abstraction of these modules. The analysis regards error injection as a deterministic process and models the injected fault as a single-bit or single-byte error, based on which both attack strategies and countermeasures are proposed.

Despite the extensive researches on fault attack analysis, there are still fundamental problems that have not been solved. The first question is how to model the fault injection and propagation process accurately. *The fault attack process is in nature probabilistic due to the following two reasons.* First, a wide range of fault injection attack techniques have been developed, which have different capability to inject faults at targeted time, denoted as temporal accuracy, and cycle-to-cycle technique parameter variation. Second, the attack strategies can also be deterministic or probabilistic. Capturing the uncertainty in attack process is important but is usually ignored in existing works.

Besides the requirement on fault modeling, the second problem is how to evaluate the system vulnerability efficiently. Because the system state space grows exponentially with the increase of system size, construction of the system FSM or calculation of state transition probability quickly become intractable even for small circuits. To avoid the exploration of FSM, different simulation-based [23] and analytical methods [32] have been proposed. Simulation-based methods provide a general approach to deal with different attack scenarios with high granularity, but may suffer from slow convergence due to large sample variance. Analytical methods [32] provide fast evaluation speed but have relatively low accuracy and granularity. How to increase the efficiency without sacrificing the accuracy is thus another critical question.

In Sect. 4.3, we propose a cross-level framework that directly targets at the two problems in fault attack evaluation. A holistic model is first proposed to capture the probabilistic nature of attack process by regarding the attack parameters as samples of random variables following different distributions. Based on the model, a security metric, defined as *System Security Factor* (SSF), is proposed to measure the vulnerability of the system and provide guidance for further design optimization. SSF is defined as the probability of a successful attack, which incorporates the uncertainty of attack process to allow for a much more accurate evaluation of different attack techniques and strategies. A cross-level evaluation framework is developed to evaluate SSF in general systems and further enhanced with a novel system pre-characterization procedure. Based on the pre-characterization, registers in the system are classified to enable a hybrid evaluation with both analytical analysis and importance Monte Carlo sampling. We summarize our contributions as follows:

- A holistic model is proposed to capture the intrinsic uncertainty of attack techniques in a probabilistic manner, based on which SSF is defined to measure the system vulnerability.
- A Monte Carlo framework is developed for SSF evaluation and further enhanced with a novel system pre-characterization procedure for better efficiency.
- Our framework is verified on a commercial processor and demonstrates good efficiency and accuracy.

The rest of Sect. 4.3 is organized as follows. In Sect. 4.3.1, we describe the motivation of our work. Section 4.3.2 formally defines the attack model, the holistic model for fault injection process, and the proposed metric on system vulnerability. In Sect. 4.3.3, we present our pre-characterization procedure, based on which an importance sampling strategy is proposed. Section 4.3.4 describes our cross-level fault modeling strategy. We demonstrate the performance of the framework in Sect. 4.3.5 and conclude our work in Sect. 4.3.6.

### 4.3.1 Motivation

Security mechanisms in modern systems are usually built in a hierarchical manner. To guarantee certain security policy, cross-level mechanisms, including hardware security primitives in circuit level, instruction set architecture support in architecture

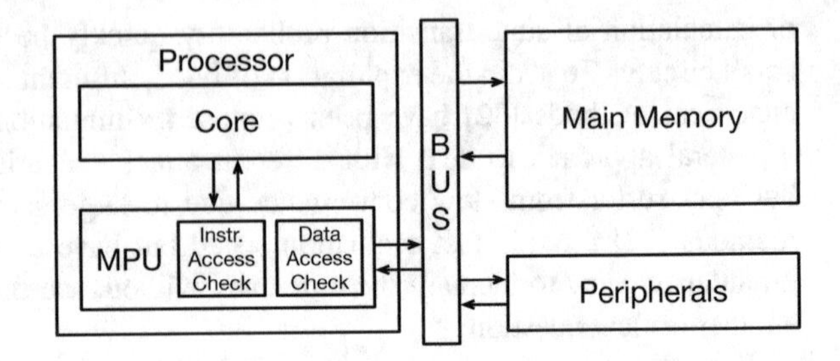


**Fig. 4.14** Diagram for MPU execution

level, firmware in application level, and so on, are combined together. Higher-level security protections rely on the correct functionality of lower-level primitives. With the security hierarchy, an end-to-end protection can be realized under cost and performance constraints.

Fault attack, however, directly targets at the lower level hardware. By injecting errors into key security primitives, the functionality of the hardware can be altered temporally, which fundamentally changes the assumption on the hardware that the entire system security mechanisms rely on. For example, in a processor, memory protection unit (MPU), as shown in Fig. 4.14, provides memory isolation features. After the data and instruction access patterns are defined, all memory access from the core and peripherals will be checked by MPU to determine whether the operations are legitimate or not. Once illegal memory operations are detected, higher-level security mechanisms will execute to isolate the process, and thus, prevent malwares from affecting other processes maliciously. If the correct behavior of an MPU is altered, it is possible for the malware to bypass the MPU and then, the entire security mechanism.

Because fault attack threatens the system security seriously, we propose our security metric and framework towards an accurate and efficient evaluation, target at helping the designers from the following aspects:

- Quantitatively characterize and compare the system vulnerability against different fault attack techniques.
- Identify security critical system components for protection under cost and overhead constraints.
- Evaluate and compare the effectiveness of different countermeasures and guide further design optimization.

In the following sections, we will formally define our security metric and describe the proposed framework.

### *4.3.2 Problem Formulation*

In this section, we formally define the proposed metric on the system vulnerability against fault attack. We first define the attack model by specifying the knowledge and target of the attackers as well as the attack flow. Then, a holistic model that captures the uncertainty in the fault injection process is proposed. The security metric is then defined based on the holistic model.

### Attack Model

We define the attack model from the following three aspects: (1) the knowledge of the attackers; (2) the target of the attackers; (3) the attack flow. The attacker's knowledge is stated as below:

- The attacker knows the system and its physical implementation.
- The attacker has physical access to the system and can inject faults into the system.
- The attacker can choose the workload program.
- The attacker does not know the system configuration.

The target of the attacker can be roughly classified into two categories: (1) bypassing the existing security mechanisms to enable malicious operations on the system [48], e.g., illegal memory access and (2) causing leakage of important system information [33], e.g., cryptographic keys.

To enable a successful attack, we consider the following unified attack flow as shown in Fig. 4.15, which is capable of representing both of the two different scenarios. We distinguish two important timing in the attack, including target cycle $T_t$ and error injection cycle $T_e$, and define $t = T_t - T_e$. Depending on the attack scenarios, the three parameters can have different meanings. For the first scenario, $T_t$ represents the time in the workload program when the operations that violate system security policies are carried out, while $T_e$ represents the time when errors are injected into the system to help the malicious operations from being detected. For the second scenario, $T_e$ still represents the time when errors are injected, while $T_t$ denotes the time when system information can be observed illegally, as stated in [8]. How to choose $T_t$, $T_e$, and $t$ highly depends on the target and attack strategies of the attackers. In Sect. 4.3, we focus on the first scenario but the proposed metric and framework are flexible for different attack categories. We assume the attackers want to access and modify certain memory locations without permission. Then, illegal memory access and modification is carried out in $T_t$, while errors are injected in $T_e$ to help with the illegal memory access.

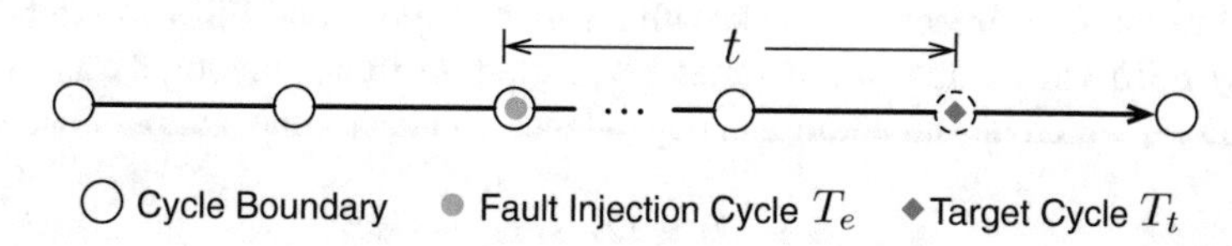


**Fig. 4.15** Fault attack flow

### Holistic Fault Injection Modeling

To inject faults into the hardware, different techniques have been proposed, including modification in power supply voltage or clock, chip overheating, focused ion beam (FIB), and so on [3]. In order to provide a generic model, we consider timing distance $t$ and technique parameters $\boldsymbol{p}$ to characterize attack process. $\boldsymbol{p}$ is a vector of characterization parameters that may vary depending on the attack techniques. For example, for attacks based on clock modification, $\boldsymbol{p}$ consists of the amplitude and duration of injected clock glitches, region impacted by the injection, and so on, while for radiation-based attacks, $\boldsymbol{p}$ includes the number of impacted cycles, location and area of the radiated spot, and so on. Because intrinsic uncertainty exists in fault injection process, we regard both $t$ and $\boldsymbol{p}$ as samples from random variable $T$ and $\boldsymbol{P}$ that follow the distribution $f_{T,\boldsymbol{P}}$. $f_{T,\boldsymbol{P}}$ is determined by considering the uncertainty of both different attack techniques and attack strategies.

In Sect. 4.3, we consider fault injection with radiation-based techniques. Because the physical mechanism of radiation-based techniques is similar to that of soft error induced by partial strike [3], we leverage the physical model following [23]. To characterize the fault injection process, we consider $\boldsymbol{p} = [g, r]$, where $g$ denotes the center of the radiation and $r$ denotes the radius of the radiated region. We assume that only one cycle is impacted by each fault injection, but our framework can easily incorporate multi-cycle impact into the evaluation. We assume one radiation can cause voltage transients at all the gates that are in the radiated region and leverage the method in [12] to determine all the impacted gates based on $g$ and $r$. Due to the temporal accuracy and parameter variation of the attack techniques, we assume the corresponding random variable $T$ and $\boldsymbol{P}$ follow a uniform distribution with the range centered at the targeted time and expected parameter. We will investigate how the change of distribution will impact the overall system vulnerability in experimental results.

### System Security Factor

In this section, we define the security metric that measures the system vulnerability. The execution of the system can be formally abstracted as an FSM $F = \langle \Omega_I, \Omega_O, \Omega_S, \delta, \rho, s_0 \rangle$, where $\Omega_I$, $\Omega_O$, $\Omega_S$, $s_0$ denote the set of possible primary inputs (PI), outputs (PO), internal states, and initial states, respectively [23]. $\delta : \Omega_S \times \Omega_I \rightarrow \Omega_S$ is the state transition function, while $\rho : \Omega_S \times \Omega_I \rightarrow \Omega_O$ is the output function.

Due to the fault injection attack with $t$ and $\boldsymbol{p}$, the original FSM is extended to a faulty FSM with a new set of states $\Omega'_S$, state-transition function $\delta'$, and output function $\rho'$, which has an additional dependence on attack time and parameters:

$$\delta' : \Omega'_S \times \Omega_I \times \Omega_T \times \Omega_{\boldsymbol{P}} \rightarrow \Omega'_S,$$

$$\rho' : \Omega'_S \times \Omega_I \times \Omega_T \times \Omega_{\boldsymbol{P}} \rightarrow \Omega_O,$$

where $\Omega_T$ and $\Omega_{\boldsymbol{P}}$ denote the set of possible values for $T$ and $\boldsymbol{P}$, respectively.

For an error-free system, malicious operations or observations may subject to existing security mechanisms and cause security violations. Fault attack can help change the original system state transitions so that security mechanisms are not triggered. Therefore, how easy are the attackers able to create such illegal transition to bypass existing mechanisms becomes the key question to answer for system vulnerability evaluation. We define SSF as the criterion to measure the probability for such illegal transition considering the uncertainty in fault attack process. More formally, let $e$ be an indicator variable representing whether the illegal transition is created. For a given benchmark, $e$ is a function of attack parameters $t$ and $\boldsymbol{p}$. Therefore, we are able to define SSF as

$$SSF = \mathrm{e}_{T,\boldsymbol{P}}(E),$$

where $E$ represents the corresponding random variable of $e$.

To evaluate SSF, we propose to use Monte Carlo based method. Let $N$ denote the total number of fault attacks at timing distance $t_1, \ldots, t_N$. Then, an empirical finite-sample estimate of SSF becomes

$$\hat{SSF} = \frac{1}{N} \sum_{t_i, \boldsymbol{p}_i \sim f_{T,\boldsymbol{P}}} e(t_i, \boldsymbol{p}_i)$$

To analyze the convergence of the Monte Carlo framework, we rely on the weak law of large number (LLN). After $N$ samples, we have

$$\Pr\left[\left|\frac{\sum_{i=1}^{N} e(t_i, \boldsymbol{p}_i)}{N} - \mathrm{e}_{T,\boldsymbol{P}}[E]\right| \geq \epsilon\right] \leq \frac{\sigma_E^2}{N\epsilon^2},$$

where $\sigma_E^2$ represents the sample variance and $\epsilon$ denotes the empirical risk. As we can see, for fixed $\epsilon$, the convergence rate of the Monte Carlo framework is determined by the $\sigma_E^2$. To reduce the sample variance and increase the convergence rate, we propose an importance sampling strategy. The basic intuition of importance sampling is to increase the probability to sample from the cycles and the attack parameters that are more likely to result in a successful attack. Let $g_{T,\boldsymbol{P}}$ denote the actual sampling distribution, then, after $N$ samples, the SSF can be estimated by

$$\hat{SSF}' = \frac{1}{N} \sum_{t_i, \boldsymbol{p}_i \sim g_{T,\boldsymbol{P}}} \frac{f_{T,\boldsymbol{P}}(t_i, \boldsymbol{p}_i)}{g_{T,\boldsymbol{P}}(t_i, \boldsymbol{p}_i)} e(t_i, \boldsymbol{p}_i)$$

$g_{T,\boldsymbol{P}}$ is critical for importance sampling. In Sect. 4.3.3, we will propose our pre-characterization strategy to determine $g_{T,\boldsymbol{P}}$.

### 4.3.3 Importance Sampling via System Pre-characterization

In this section, we describe our novel system pre-characterization procedure and derive the sampling distribution $g_{T,\mathbf{P}}$. The system pre-characterization is carried out based on the following observations.

**Observation 1** *In a system, only a few modules are related to certain security mechanism. These modules communicate with the core by a few signals and determine the related transitions of FSM.*

We denote these modules as security-critical modules and these signals as responding signals. When malicious operations are conducted and detected by these critical modules, responding signals are set to control the transition of circuit FSM to handle the security violation. To create illegitimate transitions at the presence of malicious operation to bypass the security policy, the attackers need to either prevent the security-critical modules from setting the responding signals or prevent the responding signals notifying the processor. Because only the circuits in the fanin and fanout cones of the responding signals can potentially impact them, we just consider attacks on these parts of the circuit.

Therefore, the first step of our system pre-characterization is identify responding signals based on the system specification. Then, to get the circuit components in the fanin and fanout cones of the responding signals, we unroll the circuit netlist and traverse the unrolled netlist in a breadth-first order starting from the identified signals. Because the number of responding signals is usually small, the circuits in the fanin and fanout cones are only of a small portion of the entire netlist, which helps to reduce the overall sample space and the sample variance significantly.

**Observation 2** *Bit flips at different circuit nodes can have different correlation with the flips at responding signals.*

The observation enables us to further distinguish the gates or registers in the fanin and fanout cones of the responding signals. As we have described, because the machine state transitions are controlled by these responding signals, to create illegitimate transitions, the attackers need to inject faults to either change the value of responding signals or block their propagation to the core. Circuit nodes that have a high bit flip correlation with responding signals are more likely to accomplish the target and thus, should have a higher probability in the sampling distribution.

To evaluate the bit flip correlation, we define the switching signature for each circuit node in the fanin and fanout cones. The switching signature is a sequence of binary values that indicate whether the logic value for the circuit node switches or not at each cycle. Let a binary vector $\mathrm{ss}(g)$ denotes the switching signature for circuit node $g$. For the $i$th cycle, $\mathrm{ss}_i(g) = 1$ if the logic value switches between the $(i-1)$th cycle and the $i$th cycle and $\mathrm{ss}_i(g) = 0$ otherwise. Consider circuit node $g$ in the $i$th unrolled circuit, then, the bit flip correlation between $g$ and responding signal $rs$ is computed by

$$\text{Corr}_i(g, rs) = \frac{|\text{ss}(g)\&(\text{ss}(rs) \ll i)|}{|\text{ss}(g)|},$$

where & represents the bitwise and operation and $|\cdot|$ denotes the $\ell_1$ norm, which calculates the hamming weight of the binary vector. ss($rs$) is shifted to the left by $i$ bits because it takes $i$ cycles for the bit flips in the $i$th unrolled circuit to reach the responding signal. Note that $i \geq 0$ if $g$ is in the fanin cone of $rs$ and $i < 0$ if $g$ is in the fanout cone of $rs$.

To evaluate the bit flip correlation, in the second step of our system pre-characterization, we first do a RTL-level simulation with synthetic benchmarks and keep a record of the logic value for each register. Then, gate-level logic simulation is carried out to derive the logic value as well as switching signature for each circuit node, i.e. combinational gates. Because we are able to use fast bit-parallel calculation, the overall calculation can be very efficient. Then, starting from the responding signals, we traverse their fanin and fanout cones and calculate the bit flip correlation for each circuit node. We use the following example to illustrate the calculation of bit flip correlation in a responding signal's fanin cone.

*Example 4.3* As shown in Fig. 4.16, after the logic-level simulation with synthetic benchmark, the logic value for each circuit node is recorded. The switching signature for each node is derived accordingly. Then, the switching correlation for gate $g_1$, $g_2$, $g_3$ and responding signal $rs$ can be computed as

$$\text{Corr}_0(g_1, rs) = \frac{|00101101\&(01001101 \ll 0)|}{|00101101|} = \frac{3}{4}$$

$$\text{Corr}_0(g_2, rs) = \frac{|01100111\&(01001101 \ll 0)|}{|01100111|} = \frac{3}{5}$$

$$\text{Corr}_1(g_3, rs) = \frac{|01001111\&(01001101 \ll 1)|}{|01001111|} = \frac{2}{5}.$$

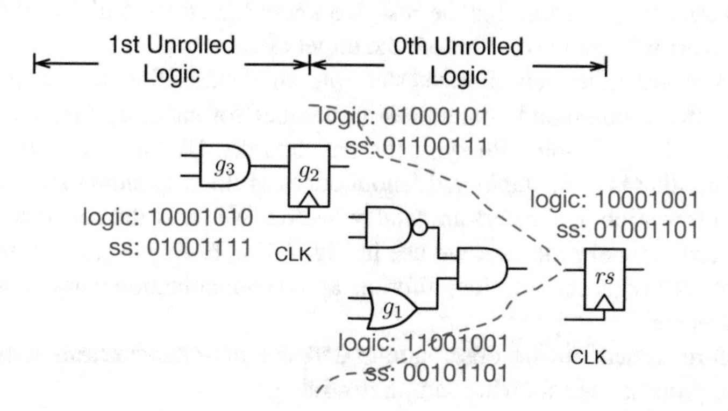


**Fig. 4.16** Example of calculating the bit flip correlation between internal gates, i.e. $g_1$, $g_2$, $g_3$, and responding signals $rs$

**Observation 3** *Bit errors can stay in some registers without propagating to other registers or getting masked.*

For these registers, sampling based method can be inefficient. This is because for bit errors injected into these registers even with a large timing distance $t$, it is still possible for them to cause illegitimate machine state transitions. Therefore, to enable an accurate evaluation, $\Omega_T$ has to be enlarged to cover a large range of values for $T$, which may lead to large sample space and large sample variance.

We denote these registers as memory-type registers and the other registers as computation-type registers. To characterize the registers, we propose two parameters: error lifetime and error contamination number. Error lifetime captures the number of cycles one bit error stays in the system before gets masked, while error contamination number captures the propagation of bit errors originating from one register to other registers. For the memory-type registers, since the bit errors tend to stay locally without further propagation or getting masked, they usually have a long error lifetime and close-to-0 error contamination number.

Because of the different characteristics of the memory-type and computation-type registers, we choose to use different strategies to evaluate the system vulnerability originated from these registers. For memory-type register, though long error lifetime makes sampling method inefficient, it actually enables us to analyze the error outcome analytically. This is because the outcome of fault attack on these registers is not determined by the timing distance between fault injection cycle and target cycle but mainly by the functionality of the memory-type registers in the system. Therefore, we choose to evaluate these registers analytically considering the system configuration, faulty registers, and benchmarks, which avoids error injection simulation without compromising the overall accuracy.

For computation-type registers, though we still use sampling based method for the evaluation, we can also increase the convergence rate due to the following two reasons. On the one hand, since the error lifetime for these registers is usually small, only a small range of $T$ need to be considered. On the other hand, after we finish sampling the timing distance $t$, we only need to consider those registers with error lifetime larger than $t$, while for the rest, we know the attack will fail because the injected errors will get masked before the target cycle.

Therefore, the third step of our system pre-characterization would be to determine the error lifetime and contamination number for each register. We leverage fast RTL-level simulation with synthetic benchmarks. Bit errors are injected into each register that is in the fanin and fanout cones of the responding signals and the two characterization parameters are then collected. We show the collected statistics from the commercial processor we use in Fig. 4.17a, b. As we can see, more than half of the total registers have long lifetime and 0 contamination number, which are classified as memory-type registers.

Therefore, based on the observations and the pre-characterization procedure above, we propose the following sampling strategy:

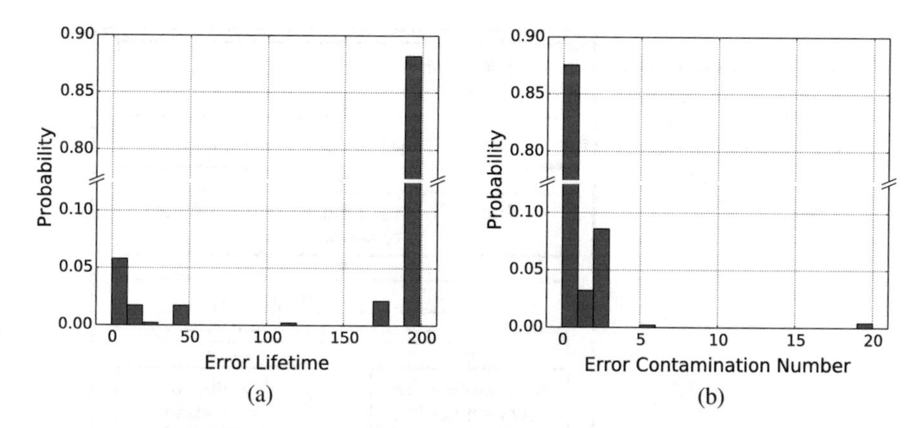


**Fig. 4.17** Distribution of characterization parameters of different registers in the commercial processor: (**a**) error lifetime and (**b**) error contamination number

- First, we decompose the overall sampling distribution $g_{T,\boldsymbol{P}}$ as

$$g_{T,\boldsymbol{P}} = g_T \cdot g_{\boldsymbol{P}|T}$$

- Then, we sample the timing distance $t$ following the distribution $g_T$. Here, for certain timing distance $t = i$, we have

$$g_T(t = i) = \frac{\omega_i}{\sum_i \omega_i},$$

where $\omega_i = \sum_{g \in \Omega_i}(1 + \alpha \mathrm{Corr}_i(g, rs)\delta(\mathrm{L}(g) \geq \beta i))$, $\Omega_i$ denotes the set of gates in the fanin and fanout cones of responding signals in the $i$th unrolled circuit and $\mathrm{L}(g)$ represents the error lifetime of $g$. If $g$ is a register, $\mathrm{L}(g)$ is the error lifetime of itself. If $g$ is a combinational gate, $\mathrm{L}(g)$ equals to the maximum error lifetime of the registers in the fanout cone of $g$ in the $i$th unrolled circuit. $\alpha$ and $\beta$ are configurable parameters that control the calculation of the distribution.
- Third, we sample the technique related parameters $\boldsymbol{p} = [g, r]$ for cycle $t = i$ following the conditional distribution $g_{\boldsymbol{P}|T}$ where

$$g_{\boldsymbol{P}|T}(\boldsymbol{p}|t = i) = \begin{cases} \frac{(1+\alpha \mathrm{Corr}_i(g,rs)\delta(\mathrm{L}(g) \geq \beta i))\mathrm{Unif}(r)}{\sum_{h \in \Omega_i} 1+\alpha \mathrm{Corr}_i(h,rs)\delta(\mathrm{L}(h) \geq \beta i)} & \text{if } g \in \Omega_i \\ 0 & \text{otherwise,} \end{cases}$$

where $\mathrm{Unif}(r)$ denotes the uniform distribution for the radius of the radiated area.

Following the procedure above, we can get one sample of $t$ and $\boldsymbol{p}$ from $g_{T,\boldsymbol{P}}$. To determine $g_{T,\boldsymbol{P}}$, only logic-level simulation is required for the three steps in our pre-characterization, which can be finished efficiently. Meanwhile, since the pre-characterization only needs to be conducted once, its introduced overhead on the overall evaluation is negligible.

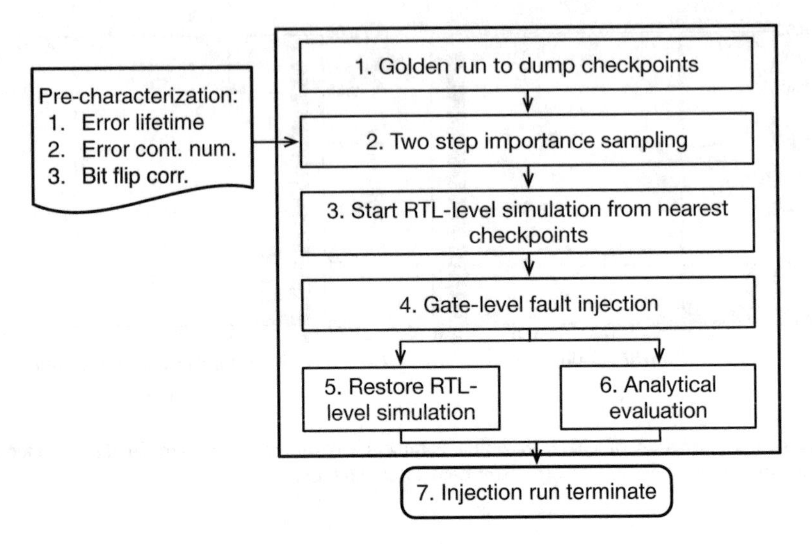


**Fig. 4.18** Overall cross-level evaluation framework

### *4.3.4 Cross-Level Fault Propagation Simulation*

In this section, we describe the proposed cross-level simulation framework. The overall flow is illustrated in Fig. 4.18. In the first step, we run the RTL-level simulation without any fault attack, denoted as golden run. During the golden run, golden checkpoints that contain the logic values of all registers are dumped. Then, we follow the importance sampling strategy to generate the sample of fault injection cycle and attack parameters. In the third step, we restart the RTL-level simulation from the nearest golden checkpoint and run the simulation to the fault injection cycle. The simulation then switches to gate level in the fourth step to evaluate the gate-level mechanisms and calculate the errors latched by registers at the end of fault injection cycle. Depending on the type of the faulty registers, we may use analytical evaluation or restore RTL-level simulation. One fault propagation simulation is terminated when the error outcome can be determined.

#### RTL-Level Golden Simulation

Before the fault attack run, a complete run of the benchmark is performed, termed as the golden run. During the golden run, golden checkpoints are dumped at intermediate points. The golden run is only performed in RTL level. Because the golden checkpoints contain the logic values of all registers in the system, they permit the following advantages:

- Golden checkpoints help reduce the warm-up simulation before the fault injection cycle in fault attack runs since the simulation can be restarted directly at the nearest golden checkpoints.
- By comparison with golden checkpoints, the outcome of fault attack run can be determined.

For one benchmark, the RTL-level golden run only needs to be conducted once.

#### RTL-Level Fault Attack Simulation

After the golden run, we first sample the timing distance $t$ and attack parameter vector $\boldsymbol{p}$ following the importance sampling strategy. Then, the RTL-level simulation is restarted from the nearest golden checkpoints and runs to the fault injection cycle. The simulation is then switched to gate level to determine the bit errors latched by registers by the end of fault injection cycle. Depending on the type of faulty registers, we choose different strategies for the downstream evaluation. When errors only exist in memory-type registers, we only need analytical evaluation to determine the error impact. Otherwise, we will inject the bit error back to RTL level to restart the RTL-level simulation. Once the RTL-level simulation resumes, the logic value of each register is dumped at different checkpoints and compared with the golden run to determine whether the target illegitimate transitions are created or not. One fault attack run is completed thereby. The whole process is continued until the empirical estimate converges.

#### Gate-Level Fault Attack Modeling

Gate-level simulation only runs for the fault injection cycle. From the sampled attack parameter vector $\boldsymbol{p}$, we can get the radiated region and determine all the voltage transients at the output of the impacted gates as described in Sect. 4.3.2. Then, as shown in Fig. 4.19a, we traverse the whole circuit netlist in a topological order to propagate the voltage transients to the registers. A voltage transient can get latched if it arrives at the register and satisfies the setup and hold time requirements as in Fig. 4.19b. Because the generated voltage transients can get latched by more than one registers, by the end of fault injection cycle, we determine a set of faulty registers.

To demonstrate the necessity of including the gate-level modeling into the overall framework, we collect the bit error patterns by the end of fault injection cycle. As shown in Fig. 4.20a, around 14.5% errors exist in multiple bytes, while for the errors within a single byte, none of the errors exists in all the bits. This proves the inaccuracy of current assumption on either single-bit or single-byte error. We also compare the number of error patterns captured by considering the attack on combinational gates and registers. As shown in Fig. 4.20b, the number of error

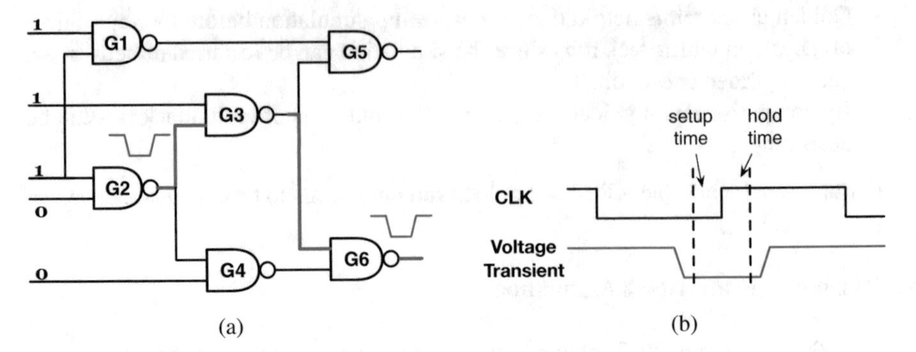


(a) (b)

**Fig. 4.19** Gate-level modeling: (**a**) propagation of voltage transients; (**b**) latching condition

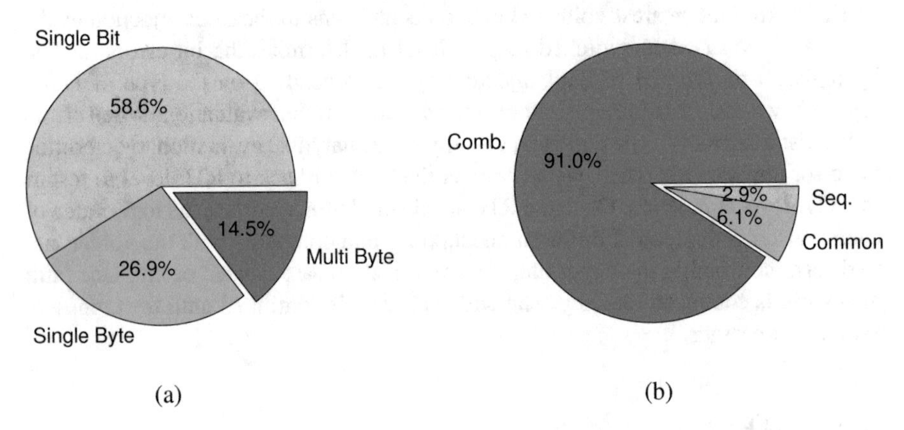


(a) (b)

**Fig. 4.20** Bit error patterns generated by gate-level simulation: (**a**) error distribution; (**b**) comparison between error patterns induced by attack on combinational gates and sequential elements

patterns induced by the attack on combinational gates is much larger than that of registers, which proves that simply considering fault attack on sequential elements, i.e. registers, is not accurate enough.

### *4.3.5 Experimental Results*

In this section, we report on the extensive experiments to demonstrate the effectiveness of the proposed Monte Carlo evaluation framework. All the experimental results are carried out on a commercial processor. We evaluate the security policy related to the memory access pattern for both core and peripherals and study the fault attack vulnerability associated with MPU. We follow the attack flow as discussed in Sect. 4.3.2 and evaluate SSF as the probability for the attacker to bypass the

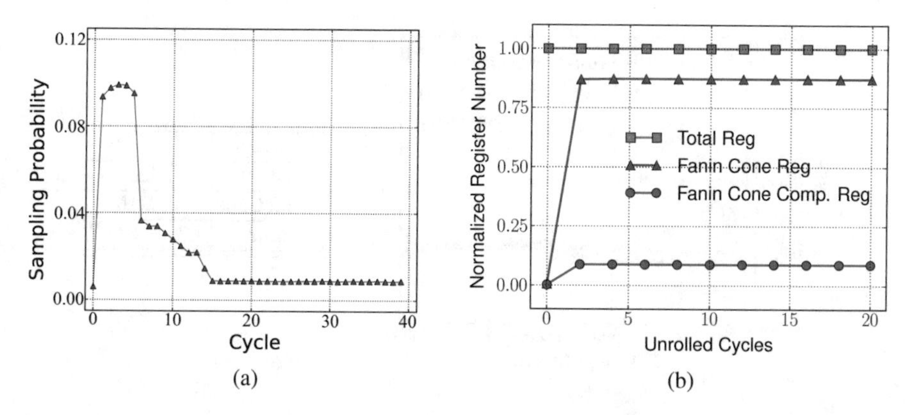


**Fig. 4.21** Effectiveness of importance sampling strategy: (**a**) sampling distribution for different value in $\Omega_T$; (**b**) reduction of sample space with our importance sampling strategy

memory access policy. The benchmark we use is written in C++ which includes illegal memory write and read operations. We use Synopsys VCS as the RTL-level simulator and implement our own gate-level simulator with C++ following the algorithm shown in [23].

We first demonstrate the effectiveness of the importance sampling strategy. The sampling distribution for the timing distance $t$ is shown in Fig. 4.21a. In Fig. 4.21b, we demonstrate the reduction of sample space. As we can see, with the importance sampling strategy, the sample space is reduced significantly. To demonstrate the increase of the convergence rate, we compare random sampling, importance sampling from fanin/fanout cones, and our mixed strategy that combines importance sampling with analytical analysis for memory-type registers. The range of $t$ is 50 cycles and the range for $P$ includes a sub-block of gates of around $1/8$ of MPU identified following [12]. As shown in Fig. 4.22b, our importance sampling strategy permits much faster convergence rate compared to the other two methods. We list the detailed statistics in Fig. 4.22. As we can see, the sample variance of our importance sampling strategy is $9.70 \times 10^{-05}$. Compared to 0.0261 of random sampling, it permits more than 2500× increase. Although the speedup is related to the systems, benchmarks, and uncertainty of attack process, the increase of convergence rate significantly increases the potential of our framework to deal with large practical systems.

From the SSF evaluation, we are able to recognize that there are around 3% registers that contribute to more than 95% SSF. Since we consider radiation-based attacks, the physical mechanisms of which are similar to that of soft error induced by particle strike [3], we are able to leverage similar protection techniques as proposed in [31, 51]. Suppose we use error resilient designs for the identified 3% registers, which permits around 10× better resilience with 3× area overhead [31, 51], then the overall SSF can be reduced by up to 6.5× with less than 2% increase of MPU area. Though the results depend on the attack and mitigation techniques, they can demonstrate the capability of our framework to identify key circuit components

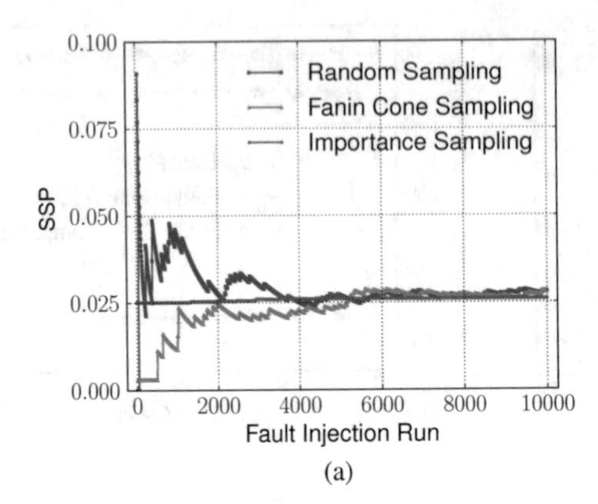


(a)

| Strategy | $s^2$ |
|---|---|
| Random | 0.0261 |
| Fanin Cone | 0.0210 |
| Our | $9.70 \times 10^{-05}$ |

(b)

**Fig. 4.22** Convergence comparison of different sampling strategies: (**a**) convergence plot; (**b**) detailed statistics for different strategies, including the number of successful attacks out of 2000 attacks and sample variance

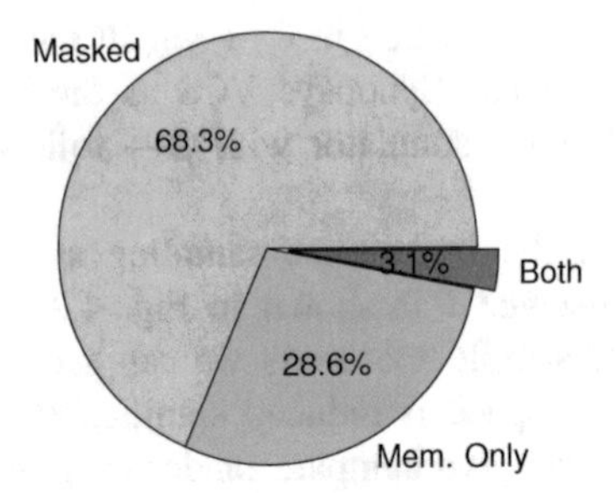


| Strategy | # Succ. Attack | SSF |
|---|---|---|
| Registers | 271 | 0.027 |
| Comb. Gates | 70 | 0.007 |

**Fig. 4.23** Comparison between SSF induced by attack on combinational gates and registers: (**a**) error statistics induced by attacking combinational gates; (**b**) SSF comparison

that are critical to the overall system security, which can be used to guide design optimization.

Then, we show the statistics of fault attacks on combinational gates. In Fig. 4.23a, as we can see, 68.4% of the fault attacks are masked. 28.6% of the attacks only cause bit errors at memory-type registers, which only requires analytical analysis. Only 3.1% of the total fault attack runs require further RTL-level simulation. This further proves the effectiveness to distinguish between memory- and computation-type registers. We also compare the SSF induced by attacks on combinational gates and registers. As shown in Fig. 4.23b, SSF due to fault attacks on combinational gates is around 25.8% of the SSF induced by attacks on registers. We also notice that all these successful attacks are from the fault injection into the fanin cone of the 3% registers as we have identified above. Therefore, to protect the MPU against fault attack, both the registers themselves and gates in their fanin cone need to be protected.

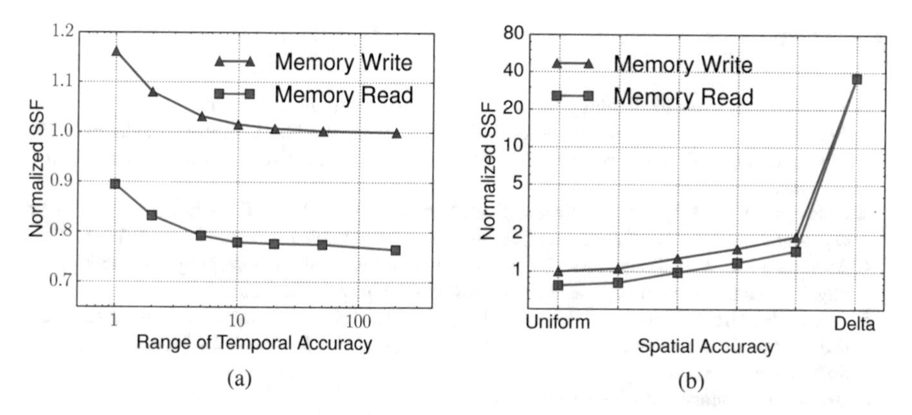


**Fig. 4.24** Impact of temporal accuracy and attack parameter variation on SSF

We also evaluate the impact of the temporal accuracy and parameter variation of the attack techniques on the overall SSF. For fixed attack parameter variation, we assume the temporal accuracy follows a uniform distribution within a specific range. As shown in Fig. 4.24a, with the overall range decreasing, the normalized SSF increases significantly for both benchmarks on illegal memory write and read. To evaluate the impact of parameter variation, we fix the temporal accuracy. As shown in Fig. 4.24b, as the variation of radiated gates changes from the largest, i.e. uniform distribution over all the gates, to the smallest, i.e. delta function centered at target gates, the normalized SSF also increases significantly. As indicated by the results, different attack techniques can have very different impact on the overall system security. The capability to capture the intrinsic uncertainty in attack process is thus important to enable a practical evaluation for different attack techniques and a feasible guidance for different systems.

### *4.3.6 Summary*

In Sect. 4.3, we propose an accurate and efficient framework to evaluate system vulnerability due to fault attack. A holistic model is proposed to capture the uncertainty of fault injection process, based on which a security metric named as SSF is defined. To evaluate SSF in practical systems, a Monte Carlo framework is proposed and further enhanced by an importance sampling strategy based on a novel system pre-characterization. A cross-level fault propagation modeling strategy that is fully compatible with the Monte Carlo simulation is also integrated. Experimental results demonstrate that with importance sampling strategy, the convergence rate is increased by up to 2500×. We also examine the impact of temporal accuracy and attack parameter variation to prove the necessity of considering the intrinsic uncertainty in attack process.

## References


1. 32nm LP Predictive Technology Model ver. 2.1 (2008). http://ptm.asu.edu
2. Agrawal, D., Archambeault, B., Rao, J. R., & Rohatgi, P. (2002). The EM side-channels. In *Proceedings of the International Conference on Cryptographic Hardware and Embedded Systems*.
3. Barenghi, A., Breveglieri, L., Koren, I., & Naccache, D. (2012). Fault injection attacks on cryptographic devices: Theory, practice, and countermeasures. In *Proceedings of the IEEE*.
4. Beckmann, N., & Potkonjak, M. (2009). Hardware-based public-key cryptography with public physically unclonable functions. In *Information Hiding* (pp. 206–220).
5. Biham, E., Granboulan, L., & Nguyen, P. Q. (2005). Impossible fault analysis of RC4 and differential fault analysis of RC4. In *Proceedings of the International Workshop on Fast Software Encryption*.
6. Boost C++ Library. http://www.boost.org
7. Chasta, N. K. (2013). A very high speed, high resolution current comparator design. *Journal of Electric, Electronics Science and Engineering, 7*(11), 1204–1207.
8. Chen, C.-N., & Yen, S.-M. (2003). Differential fault analysis on AES key schedule and some countermeasures. In *Proceedings of the Australasian Conference on Information Security and Privacy*.
9. Cunningham, P., & Delany, S. J. (2007). k-nearest neighbour classifiers. *Multiple Classifier Systems, 34*(8), 1–17.
10. Dinic, E. A. (1970). Algorithm for solution of a problem of maximum flow in networks with power estimation. *Soviet Mathematics. Doklady, 11*(5), 1277–1280.
11. Fan, J., Guo, X., De Mulder, E., Schaumont, P., Preneel, B., & Verbauwhede, I. (2010). State-of-the-art of secure ECC implementations: A survey on known side-channel attacks and countermeasures. In *Proceedings of the IEEE International Symposium on Hardware Oriented Security and Trust*.
12. Fazeli, M., Ahmadian, S. N., Miremadi, S. G., Asadi, H., & Tahoori, M. B. (2011). Soft error rate estimation of digital circuits in the presence of multiple event transients (METs). In *Proceedings of the Design, Automation and Test in Europe*.
13. Gao, M., Lai, K., & Qu, G. (2014). A highly flexible ring oscillator PUF. In *Proceedings of the IEEE/ACM Design Automation Conference* (pp. 1–6).
14. Gassend, B., Clarke, D., Van Dijk, M., & Devadas, S. (2002). Silicon physical random functions. In *Proceedings of the ACM Conference on Computer & Communications Security* (pp. 148–160).
15. Goldberg, A. V., & Tarjan, R. E. (1988). A new approach to the maximum-flow problem. *Journal of the ACM, 35*(4), 921–940.
16. Goldschlager, L. M., Shaw, R. A., & Staples, J. (1982). The maximum flow problem is log space complete for P. *Theoretical Computer Science, 21*(1), 105–111.
17. Hemme, L. (2004). A differential fault attack against early rounds of (triple-) DES. In *Proceedings of the International Conference on Cryptographic Hardware and Embedded Systems*.
18. Hund, R., Willems, C., & Holz, T. (2013). Practical timing side channel attacks against kernel space ASLR. In *Proceedings of the IEEE Symposium on Security and Privacy*.
19. International technology roadmap for semiconductor 2014. Available: https://www.itrs.net/ [Online; accessed Nov. 2014].
20. Kelner, J. A., Lee, Y. T., Orecchia, L., & Sidford, A. (2014). An almost-linear-time algorithm for approximate max flow in undirected graphs, and its multicommodity generalizations. In *Proceedings of ACM-SIAM Symposium on Discrete Algorithms* (pp. 217–226). Philadelphia: SIAM.
21. Lee, J. W., Lim, D., Gassend, B., Suh, G. E., Van Dijk, M., & Devadas, S. (2004). A technique to build a secret key in integrated circuits for identification and authentication applications. In *Proceedings of the Symposium on VLSI Technology and Circuits* (pp. 176–179).

22. Li, M., Miao, J., Zhong, K., & Pan, D. Z. (2016). Practical public PUF enabled by solving max-flow problem on chip. In *Proceedings of the IEEE/ACM Design Automation Conference* (p. 164). New York: ACM.
23. Li, M., Wang, Y., & Orshansky, M. (2016). A Monte Carlo simulation flow for SEU analysis of sequential circuits. In *Proceedings of the IEEE/ACM Design Automation Conference*.
24. Li, Y., Sakiyama, K., Gomisawa, S., Fukunaga, T., Takahashi, J., & Ohta, K. (2010). Fault sensitivity analysis. In *Proceedings of the International Conference on Cryptographic Hardware and Embedded Systems*.
25. Lin, T.-M., & Mead, C. (1984). Signal delay in general RC networks. *IEEE Transactions on Computer-Aided Design of Integrated Circuits and Systems, 3*(4), 331–349.
26. Maiti, A., Gunreddy, V., & Schaumont, P. (2013). A systematic method to evaluate and compare the performance of physical unclonable functions. In *Embedded systems design with FPGAs* (pp. 245–267). New York: Springer.
27. Majzoobi, M., & Koushanfar, F. (2011). Time-bounded authentication of FPGAs. *IEEE Transactions on Information Forensics and Security, 6*(3), 1123–1135.
28. Mead, C., & Ismail, M. (2012). *Analog VLSI implementation of neural systems* (Vol. 80). Berlin/Heidelberg: Springer Science & Business Media.
29. Mehr, I., & Welland, D. R. (1997). A CMOS continuous-time G m-C filter for PRML read channel applications at 150 mb/s and beyond. *IEEE Journal of Solid-State Circuits, 32*(4), 499–513.
30. Miao, J., Li, M., Roy, S., & Yu, B. (2016). LRR-DPUF: Learning resilient and reliable digital physical unclonable function. In *Proceedings of the International Conference on Computer Aided Design*.
31. Mitra, S., Seifert, N., & Zhang, M. (2005). Robust system design with built-in soft-error resilience. *Journal of Computer, 38*, 43–52.
32. Mukherjee, S. S., Emer, J., & Reinhardt, S. K. (2005). The soft error problem: An architectural perspective. In *Proceedings of the International Symposium on High-Performance Computer Architecture*.
33. Nahiyan, A., Xiao, K., Yang, K., Jin, Y., Forte, D., & Tehranipoor, M. (2016). AVFSM: A framework for identifying and mitigating vulnerabilities in FSMs. In *Proceedings of the IEEE/ACM Design Automation Conference*.
34. Plotkin, M. (1960). Binary codes with specified minimum distance. *IRE Transactions on Information Theory, 6*(4), 445–450.
35. Potkonjak, M., & Goudar, V. (2014). Public physical unclonable functions. *Proceedings of IEEE, 102*(8), 1142–1156.
36. Potkonjak, M., Meguerdichian, S., Nahapetian, A., & Wei, S. (2011). Differential public physically unclonable functions: Architecture and applications. In *Proceedings of the IEEE/ACM Design Automation Conference* (pp. 242–247).
37. Rajendran, J., Rose, G. S., Karri, R., & Potkonjak, M. (2012). Nano-PPUF: A memristor-based security primitive. In *Proceedings of the IEEE Annual Symposium on VLSI* (pp. 84–87).
38. Rührmair, U. (2009). *SIMPL Systems: On a Public Key Variant of Physical Unclonable Functions*. IACR Cryptology ePrint Archive, 2009, p. 255.
39. Rührmair, U., Chen, Q., Stutzmann, M., Lugli, P., Schlichtmann, U., & Csaba, G. (2010). Towards electrical, integrated implementations of SIMPL systems. In *IFIP International Workshop on Information Security Theory and Practices* (pp. 277–292). Heidelberg: Springer.
40. Rührmair, U., Sehnke, F., Sölter, J., Dror, G., Devadas, S., & Schmidhuber, J. (2010). Modeling attacks on physical unclonable functions. In *Proceedings of the ACM Conference on Computer & Communications Security* (pp. 237–249).
41. Salmani, H., & Tehranipoor, M. (2013). Analyzing circuit vulnerability to hardware Trojan insertion at the behavioral level. In *Proceedings of the IEEE International Symposium on Defect and Fault Tolerance in VLSI and Nanotechnology Systems*.
42. Shiloach, Y., & Vishkin, U. (1982). An O(n2 log(n)) parallel max-flow algorithm. *Journal of Algorithms, 3*(2), 128–146.

43. Sun, Y., Swang, Y., & Lai, F. C. (2007). Low power high speed switched current comparator. In *Proceedings of the International Conference Mixed Design of Integrated Circuits and Systems* (pp. 305–308).
44. Suykens, J. A. K., & Vandewalle, J. (1999). Least squares support vector machine classifiers. *Neural Processing Letters, 9*(3), 293–300.
45. Tunstall, M., Mukhopadhyay, D., & Ali, S. (2011). Differential fault analysis of the advanced encryption standard using a single fault. In *Proceedings of the International Workshop on Information Security Theory and Practices*.
46. Van Woudenberg, J.G.J., Witteman, M. F., & Menarini, F. (2011). Practical optical fault injection on secure microcontrollers. In *Proceedings of the IEEE Workshop Fault Diagnosis and Tolerance in Cryptography*.
47. Yoo, A., Chow, E., Henderson, K., McLendon, W., Hendrickson, B., & Çatalyürek, Ü. (2005). A scalable distributed parallel breadth-first search algorithm on BlueGene/L. In *Proceedings of the International Conference for High Performance Computing, Networking, Storage and Analysis* (p. 25).
48. Yuce, B., Ghalaty, N. F., Deshpande, C., Patrick, C., Nazhandali, L., & Schaumont, P. (2016). FAME: Fault-attack aware microprocessor extensions for hardware fault detection and software fault response. *Proceedings of the International Workshop on Hardware and Architectural Support for Security and Privacy*.
49. Yuce, B., Ghalaty, N. F., & Schaumont, P. (2015). Improving fault attacks on embedded software using RISC pipeline characterization. In *Proceedings of the IEEE Workshop Fault Diagnosis and Tolerance in Cryptography* (2015).
50. Yuce, B., Ghalaty, N. F., & Schaumont, P. (2015). TVVF: Estimating the vulnerability of hardware cryptosystems against timing violation attacks. In *Proceedings of the IEEE International Symposium on Hardware Oriented Security and Trust*.
51. Zhang, M., Mitra, S., Mak, T. M., Seifert, N., Wang, N. J., Shi, Q., et al. (2006). Sequential element design with built-in soft error resilience. *IEEE Transactions on Very Large Scale Integration (VLSI) Systems*

# Chapter 5
# Conclusion and Future Work

This book has proposed a synergistic framework for hardware IP privacy and integrity protection. The five proposed algorithms are all built upon rigorous mathematical modeling and can collaborate with each other to prevent cross-stage violations. Our split manufacturing algorithm can harness the fabrication technology advances for Trojan prevention, which achieves a much better guarantee on the security and efficiency. The importance of IP protection against reverse engineering is recognized by both leading industrial companies, e.g., Mentor Graphics, and academia. Our AND-tree based camouflaging strategy achieves provably secure protection against the SAT-based attack and our TimingSAT algorithm enables efficient security evaluation for all the existing protection schemes. To prevent fault injection attacks, our fault attack evaluation framework provides a fast security analysis to identify security-critical system components and compare different protection strategies. The framework is prototyped by a leading IP vendor, Arm Inc., and applied on commercial products. The proposed PPUF design achieves a formal guarantee on the ESG with different design techniques proposed to improve the practicality of the design. The proposed design serves as a novel security primitive that can support public cryptography.

With the further globalization of the semiconductor supply chain, the IP privacy and integrity protection is expected to become more and more important. Hence, we anticipate more research efforts from the following perspectives.

- Split manufacturing friendly placement strategy. Existing split manufacturing strategies for Trojan prevention mainly focus on the gate-level netlist. Although different placement refinement techniques have been proposed, they usually incur large overhead. To reduce the overhead and enhance the practicality of split manufacturing, new placement strategies are required, especially considering the emerging fabrication technology, including 3D integration and so on.


M. Li, D. Z. Pan, *A Synergistic Framework for Hardware IP Privacy and Integrity Protection*, https://doi.org/10.1007/978-3-030-41247-0_5

- Full chip IC camouflaging strategy. Most of the existing IC camouflaging algorithms focus on combinational logic while the researches on the full system level and the system-on-chip (SoC) level are still missing. Compared with the protection of combinational logic, system level and SoC level protection are much more challenging. On one hand, SoC designs are much larger than combinational circuits. Hence, the scalability of the existing camouflaging and de-camouflaging algorithms can be a problem. While existing algorithms heavily rely on partitioning, its implication on the overall security is unclear. On the other hand, different modules and components in a SoC have different functionalities and structural characteristics, which may impact the strategies that can be applied to them. Moreover, useful SoC level benchmarks are also unavailable to fully support the research on camouflaging and de-camouflaging.
- Generic and quantitative security analysis for IC camouflaging. Although quantitative security analysis against exact attacks is proposed in the dissertation, how to evaluate the security against approximate attacks is still an open question and has to rely on empirical evaluation. However, such empirical evaluation is insufficient to provide formal security guarantee and suffers from an over-estimation of the provided security level. To enable generic and quantitative security analysis, cryptographic criteria can be a promising direction since they usually enjoy a high level formalism. How to bridge these cryptographic criteria and IC camouflaging is an important direction to explore.
- PPUF design with provable security against model-building attacks. The proposed PPUF achieves formally justified ESG. However, the security against model-building attacks is only evaluated empirically with parametric and non-parametric attack strategies. In fact, there are no PUF designs proposed so far with a formal security guarantee. An important direction, therefore, is how to analyze the resilience against the model building attack and how to design new PPUF with a formal guarantee against the model-building attack.

# Index

**A**
Active learning, 43–46, 48, 74
Analytical camouflaging security analysis, 72, 117, 124
AND-tree
  camouflaging strategy, 69–72
  decomposability characterization, 57–58
  general circuits, 54–55
  input bias, 55–57
  original netlist detection, 59–62
  removal attack, 64–66
  security analysis, 52–54
  stochastic greedy insertion, 62–64

**B**
Back-end-of-line (BEOL)
  FEOL, 3
  globalized supply chain, 2
  interconnections, 9
  iterative strategy, 23
  nodes, 35
  security, 29

**C**
Challenge-response pairs (CRPs), 4, 99, 100, 109, 111, 112

**E**
Execution and simulation (ESG)
  amplification, 109
  computation algorithms, 100
  lower bound, 106–108
  PPUFs, 4, 98, 103–106, 136
  simulation model, 5, 112–114
  upper bound, 108–109
  verifier's task, 109

**F**
Fault attack
  cross-level framework, 117
  cryptographic modules, 97
  flow, 119
  gate-level, 127–128
  IP integrity issues, 4
  PPUF design, 5
  RTL-level, 127
  SSF, 121
  system vulnerability, 4, 116
  valuation framework, 135
Finite state machine (FSM), 116, 117, 120, 122
Full chip camouflaging, 136

**H**
Hardware Trojan
  BEOL, 2
  fabrication time, 9
  FEOL, 2
  globalized supply chain, 1, 2
  insertion, 17
  Moore's law, 1
  PUF and PPUF, 4
  split manufacturing, 3


M. Li, D. Z. Pan, *A Synergistic Framework for Hardware IP Privacy and Integrity Protection*, https://doi.org/10.1007/978-3-030-41247-0

## I

IC camouflaging technique
  active learning, 44–45
  AND-tree (*see* AND-tree)
  arms race evolution, 40–42
  cell design
    discussion, 50–52
    generation strategy, 68–69
    STF-type, 49–50
    XOR-type, 48–49
  contributions, 43–44
  effectiveness, 74
  fabrication-level techniques, 3, 39
  SAT-based attack strategies, 5, 40
  security
    analysis, 45–48
    level, 43
  state-of-the-art techniques, 66–67
  structural and functional impact, 72–74
Importance sampling
  analytical analysis, 129
  attack parameters, 121
  effectiveness, 129
  gate-level, 127–128
  RTL-level
    fault attack simulation, 127
    golden simulation, 126–127
  strategy, 117
  system pre-characterization, 122–126, 131
Intellectual property (IP) design
  globalization, 135
  integrity
    issues, 4
    and privacy, 1–4
  privacy, 39
  protection, 135
  violations, 1, 3
IP privacy and integrity violations
  hardware, 1–4
  reverse engineering, 39
  semiconductor supply chain, 135

## K

k-isomorphism, 18–20

## L

Lagrangian relaxation algorithm
  minimum-cost flow transformation, 26–28
  multiplier update, 28–29

## M

Mixed-integer linear programming (MILP) formulation
  coefficients, 36
  dummy wires and gates, 21
  FEOL layer generation, 10, 21–25
  LR-based algorithms, 32
  minimum-cost flow transformation, 28, 31
Monte Carlo evaluation framework
  experimental results, 128–131
  FSM, 116
  motivation, 117–118
  problem formulation
    attack model, 119
    holistic fault injection modeling, 120
    system security factor, 120–121
  SSF, 117

## P

PPUF design
  basic building block, 103–106
  crossbar
    placement, 111
    structure, 110–111
  ESG, 98, 99, 109
  execution and simulation, 98
  experimental results
    ESG, 112–114
    simulation model, 112–114
  input challenge pre-pruning, 111–112
  lower bound, 106–108
  Max-flow problem, 100–102
  model-building attack resilience, 115
  MOS transistors, 98
  protocol, 99–100
  signal delay, 102–103
  statistical evaluation, 114–115
  topology, 103–106
  upper bound, 108–109
  verification, 114
  verifier's task, 109
Provably secure
  IC camouflaging (*see* IC camouflaging technique)
  SAT-based reverse engineering attack, 5

## R

Register-transfer level (RTL), 123, 126–127, 129, 130

Reverse engineering
arms race, 3, 41
black-box functional circuit, 80
camouflaging
cells, 39
connections, 40
de-camouflaging iterations, 91
FEOL and BEOL, 9
IP design (*see* Intellectual property (IP) design)
wave-pipelining, 77

**S**

SAT-based attack
active learning, 48
camouflaging cells, 76
clique-based obfuscation, 42
de-camouflaging complexity, 54
security guarantee, 40
timingSAT algorithm, 135
Simultaneous cell and wire insertion, 10, 12, 21
Split manufacturing
attack model of untrusted foundries, 11
dummy cells, 10
experimental setup, 31
fabrication technology, 135
FEOL
generation strategy comparison, 31–34
layers, 9
friendly physical design, 135
k-security realization, 18–20
motivation, 11–12
overhead and framework parameters, 36
physical proximity examination, 34–36
refinement technique, 10
security analysis, 13–18
state-of-the-art, 12–13
Statistical attack space, 5, 112, 114, 115
System security factor (SSF)
accuracy and attack parameter, 131
comparison, 130
illegal transition, 121
pre-characterization procedure, 117
radiation-based attacks, 129
system vulnerability, 117

**T**

Timing-based camouflaging, 76–77
high-entropy, 93
motivating example, 78–79
traditional strategy, 93–94
TimingSAT
algorithm, 135
attack efficiency, 42
discussion, 87
efficiency, 88
input query, 81–84
key post-processing, 86–87
netlist simplification, 84–86
performance, 89
runtime dependency, 88–90
scalability issue, 75
TU insertion, 80–81
unrolling time frames, 90–92
Trojan prevention
hardware, 10
insertion, 9
k-secure layout refinement, 29–30
Lagrangian algorithm (*see* Lagrangian relaxation algorithm)
MILP-based FEOL generation, 21–25
security and efficiency, 135

Printed in the United States
by Baker & Taylor Publisher Services